DARWINIAN MISADVENTURES IN THE HUMANITIES

DARWINIAN MISADVENTURES
IN THE
HUMANITIES

EUGENE GOODHEART

TRANSACTION PUBLISHERS
NEW BRUNSWICK (U.S.A.) AND LONDON (U.K.)

Copyright © 2007 by Transaction Publishers, New Brunswick, New Jersey.

All rights reserved under International and Pan-American Copyright Conventions. No part of this book may be reproduced or transmitted in any form or by any means, electronic or mechanical, including photocopy, recording, or any information storage and retrieval system, without prior permission in writing from the publisher. All inquiries should be addressed to Transaction Publishers, Rutgers—The State University of New Jersey, 35 Berrue Circle, Piscataway, New Jersey 08854-8042. www.transactionpub.com

This book is printed on acid-free paper that meets the American National Standard for Permanence of Paper for Printed Library Materials.

Library of Congress Catalog Number: 2007021242
ISBN: 978-1-4128-0661-9
Printed in the United States of America

Library of Congress Cataloging-in-Publication Data

Goodheart, Eugene.
 Darwinian misadventures in the humanities / Eugene Goodheart.
   p. cm.
  Includes index.
  ISBN 978-1-4128-0661-9 (alk. paper)
  1. Humanities. I. Title.

AZ101.G66 2007
001.3—dc22   2007021242

Contents

Acknowledgements	vii
Prologue	1
1. Reducing Literature and the Arts	11
2. Demystifying Religion	27
3. Reinventing Ethics	51
4. Is History a Science?	65
5. Condescending to Science	79
6. In Defense of Dualism	89
Epilogue	109
Works Cited	121
Index	125

Acknowledgements

The idea for this book grew out of a faculty seminar in Consilience at Brandeis University in which I participated as a dissident for two years. I am grateful to my colleagues, Robin Miller and Irving Epstein who organized the seminar. An early draft, so to speak, of the book appeared in a short article in *The Common Review* (fall 2002) under the title, "Imperial Science". Early versions of the chapters "Reinventing Ethics" and "Is History a Science?" (the former under the title, "Peter Singer's Challenge") appeared in *Philosophy and Literature* in October 2005 and April 2006 respectively. I am grateful to the editors of the journals for permission to reprint whatever of the original passages in the essays I retained in writing this book. Irving Louis Horowitz helped enormously by pointing out a misdirection in an earlier version of the manuscript. My anthropologist wife Joan Bamberger's informed reading of my discussion of religion gave me confidence in the argument I was making.

Prologue

The heyday of postmodern theory in the humanities has passed, though a deep skepticism of institutions and texts persists, particularly in literary studies. "Reality" and "truth," always enclosed in quotation marks, are under suspicion as counters for false authority and false consciousness. As a pervasive attitude, radical skepticism in practice is exhausting and almost impossible to sustain. In the humanities, it has had a hollowing out effect. So it should not come as a complete surprise that exponents of a theory confident about its capacity to represent reality should try to rush in to fill the vacuum. I have in mind sociobiologists and evolutionary psychologists, the current masters of the neo-Darwinian synthesis, who seem prepared to extend their domain to the humanities. I hold no brief for their overreaching, but I would be less than candid not to acknowledge that the present condition of the humanities (its self-doubt and disarray) is a kind of invitation to them. As the philosopher Thomas Nagel prophesied in his essay "The Sleep of Reason," "[Postmodernism] may now be on the way out, but I suspect that there will continue to be a market in the huge American academy for a quick fix of some kind. If it is not social constructionism, it will be something else—Darwinian explanations of practically every thing else" (Patai and Corral, 552). In previous work I have addressed the threats that radical skepticism and ideology critique pose to the integrity of the humanities. Now I turn to a threat from an entirely different quarter.

E. O. Wilson's *Consilience: The Unity of Knowledge* (1998) promotes his neo-Darwinian theory of sociobiology as the master discipline for all the disciplines. Wilson is a world-renowned Harvard biologist, who in his long and distinguished career has written about everything from the social lives of ants to the planet's fragile network of organisms. His arguments on behalf of biodiversity have earned him well-deserved praise from environmentalists. Sociobiology in its essence is a theory that promotes the view that our genetic structure largely shapes, if it doesn't absolutely determine, individual and social life. It is the antithesis to social constructionism in its denial of any significant role to genetics in the shaping of human life. Sociobiology has its critics both within and outside the discipline of biology for what they regard as its excessive emphasis on the role of genes in the individual and social development of human beings. As we might expect, the theory is anathema to humanists of postmodernist persuasion who subscribe to social constructionism. They view sociobiology as a reactionary justification of the capitalist status quo and, even more seriously, as a possible rationale for racism. The critics within the discipline of biology, very much in the minority, also view the theory as reactionary and racist, though, as biologists they cannot accept the social constructionist denial of the biological factor. As will become clear, my quarrel with sociobiology has little in common with either the social constructionists or the critics within the discipline.

Sociobiology does not exclude the role of environment and culture in evolution, but the influence of genetics is preponderant. Wilson speaks of anthropology as "breaking into two cultures of its own, different but equal (of course) in merit." There are, on the one hand, "the biological anthropologists, who attempt to explain culture as ultimately a product of the genetic history of humanity, renewed each generation by the decisions of individuals influenced by that history," and on the other, "the cultural anthropologists, [who] see culture as a higher order phenomenon largely free of genetic history and diverging from one society to the next virtually without limit" (202). The respectful nod to cultural anthropology may be well intentioned, but Wilson's sympathies are with the biological anthropologists, and he sees no immediate prospect of overcoming the schism between the two camps. His allegiance is to "the genetic history of humanity."

In *The Selfish Gene*, Richard Dawkins has tried to fill what has been a lacuna in sociobiological theory. He has introduced the *meme* as the cultural counterpart to the gene. "Mimeme comes from a suitable Greek root, but I want a monosyllable that a sounds a bit like 'gene'...it could alternately be thought of as relating to 'memory' or to the French work *meme*" (192). Though Dawkins does not specify its source or location in the brain, this agent of culture remains squarely within biological discourse. Memes are comparable to "ideas," but not as the philosopher Daniel Dennett notes in *Consciousness Explained*, in the sense of the "simple ideas" of Locke and Hume. They have a physical embodiment such as "the wheel," "calculus," "chess," "impressionism," "*The Odyssey*." According to Dawkins, the meme "is achieving evolutionary change at a rate which leaves the old gene panting far behind." Which is to say that cultural change proceeds at a faster pace than biological change. Memes travel and replicate from brain to brain. "Some of these memes are 'good' perhaps [e.g., chess], and others 'bad' [e.g. conspiracy theory]; what they have in common is a phenotypic effect that tends to disable the selective forces arrayed against them." Dennett cites the examples of "conspiracy theory" that persists "independently of their truth" and "faith" that survives in a rationalistic environment. What are those "selective forces" that have the capacity to resist and discriminate among memes? They are themselves memes "for normative concepts—for *ought* and *good* and *truth* and *beauty*" (298).

The meme is the product of speculation, not experimental discovery. Like the gene, it replicates itself. What is not clear is how it produces cultural variation. Impressionism, for instance, differs in the hands of Renoir, Pissarro, and Monet. Within a particular artistic or intellectual tradition (or a meme, as Dawkins would have it) changes and variations occur as they pass from one mind to another. John Casti believes that Dawkins has, in effect, made an argument against sociobiology: "…it has always seemed to me that his notion of a cultural *meme* playing the role for cultural traits that genes play for physiological ones is really a statement against the genetically based claims of the mainline sociobiologists" (105). On the contrary, the meme, like the gene, has its physiological source in the brain. There is epistemological harmony between meme and gene. In any event, the conception of the role of culture in the neo-Darwinian synthesis is rudimentary.

Whatever its shortcomings in the understanding of cultural change, sociobiology has established itself as the core of the prevailing neo-Darwinian synthesis. The field of biology has made extraordinary, indeed, world-transforming strides in recent decades. The discovery of DNA had already occurred and the discovery of the human genome was in prospect at the time of the writing of *Consilience*, so it is understandable that Wilson, a preeminent master of the field, would look upon the ascendancy of postmodern radical skepticism with dismay. He writes: "Postmodernism is the ultimate polar antithesis to the Enlightenment. The difference between the two extremes can be explained as follows: Enlightenment thinkers believe we can know everything, and radical postmodernists believe we can know nothing" (44). Wilson is clearly on the side of the Enlightenment, or at least his version of it. He writes with supreme confidence that biology is on the verge of possessing solutions to many of life's problems. His quarrel with "know nothing" postmodernism, deserving of sympathy, has however led him and his followers to the extreme and untenable view that everything can be known scientifically, and in particular by evolutionary biologists and psychologists. It is the extremism of this view and its frequently absurd consequences for study in fields outside the natural sciences that is the subject of this study. Believing as he does that the disciplines of the humanities and the social sciences in their current incarnations lack the scientific integrity that would justify their existence, he proposes to redress the condition by providing each discipline with a basis in biology. As we shall see, he is not alone in the enterprise.

Wilson has appropriated, some would say misappropriated, the term consilience from the nineteenth-century philosopher of science William Whewell, who also coined the term "scientist" and who, incidentally, was hostile to Darwinism. As Stephen J. Gould has pointed out in *The Hedgehog, the Fox and the Magister Pox*, Whewell means something quite different from Wilson's use of the term, a non-reductionist "'jumping together' of otherwise disparate facts into a unitary explanation" (192). Gould, a critic of sociobiology, "finds nothing either viscerally or intellectually appealing" in the desire for unification through reductionism. He proposes the alternative of "the consilience of equal regard," by which he means the kind of mutual respect among disciplines lacking in Wilson's conception. By contrast, for Wilson consilience or the unity of knowledge is the *reductionist* hierarchy that exists in the natural sciences.

> To make any progress [researchers] must meditate on the networks of cause and effect across adjacent levels or organization—from subatomic particles to atoms, say, or organisms to species—and they must think on the hidden design and forces of the networks of causation. Quantum physics thus blends into chemical physics, which explains atomic bonding and chemical reactions, which form the foundation of molecular biology, which demystifies cell biology (Wilson, 59).

Despite Wilson's use of the word "demystifies" to describe the relationship between molecular and cell biology, hierarchy in the natural sciences does not involve "replacing one field of knowledge by another," but rather consists of "*connecting* or *unifying* them" (Pinker, 70). In joining the humanities to the natural sciences, however, Wilson means to replace the present practices of the humanistic disciplines. If Wilson has his way, the humanities and the social sciences will find their place in a hierarchy with biology at the base where the practitioners of these disciplines will acquire knowledge that had eluded its present practices.

Wilson regards the humanities in their present incarnation as mystifications of their subject matter. "Philosophy, the contemplation of the unknown, is a shrinking dominion. We have the common goal of turning as much philosophy as possible into science" (12). According to Wilson, philosophy has effectively become obsolete in its understanding of mental activity and should yield its claim to knowledge about mind to cognitive and neuroscientists. The reflections of philosophers from Descartes and Kant proceed from introspection and simply draw us away from the actual operations of the brain, which is essentially "a machine assembled not to understand itself, but to survive" (105). (Why self-understanding and survival are mutually exclusive is by no means clear.) Religion too must yield to science: "Could Holy Writ be just the first literate attempt to explain the universe and make ourselves significant within it? Perhaps science is a continuation on new and better-tested ground to attain the same end. If so, then in that sense science is religion liberated and writ large" (6-7). (Perhaps not, if religion is understood not simply as an attempt to understand the origins of the material universe, but rather as a pursuit of spiritual fulfillment.) On Wilson's view, literary theory too becomes the domain of science. Literary theorists must give way to evolutionary psychologists who have formulated the rules that will explain the emergence of genius and creative achievement. "Human nature," Wilson writes, is the set of "epigenetic rules, the hereditary regularities of mental development that bias cultural evolution in one direction as opposed to another and thus connect the genes to culture" (178). His conception of culture as a genetically determined epiphenomenon is analogous to the vulgar Marxian conception of culture as a superstructure of economic reality. At a seminar at Brandeis University (November 13, 2000), which I attended, devoted to the subject of consilience, a participant asked Wilson whether knowledge of the epigenetic rules that govern art (assuming they exist) would be of value to a creative artist. The questioner assumed that he would answer "no" and was surprised that the answer was "yes." The follow-up question was whether the knowledge of the rules and their application would not prove to be an obstacle to creative originality to which the answer was also in the affirmative. Revealingly, in the course of the exchange, Wilson mentioned as evidence for a rule-bound view of the creative imagination the formulas that guide the writing of scripts for the cinema and television. Here is a case of a diminishing reductionism in the very conception of art. These formulas hardly have the dignity of *creative* universals.

Reductionism is a view to which most natural scientists subscribe. *In Facing Up: Science and Its Cultural Adversaries*, the physicist Steven Weinberg distinguishes between "grand reductionism" and "petty reductionism." The "petty" variety explains things "by studying their constituents," that is, by analyzing or reducing wholes to their parts. "[A] diamond is hard because the carbon atoms of which it is composed can fit together neatly" (111). Grand reductionism, the more significant for the theory and practice of science, is "the view that all of nature is the way it is (with certain qualifications about initial conditions and historical accidents) because of simple universal laws, to which all other scientific laws may in a sense be reduced." Within biology, the laws of the molecule are more fundamental than those of the cell. Within the constellation of the sciences, the laws of physics are fundamental. Which is not to say that reductionism is the sole mode of operation in the sciences. "As we look at nature at levels of greater and greater complexity, we see phenomena emerging that have no counterpart at the simpler levels, least of all at the level of elementary particles" (Wineberg, *Dreams*, 39). As Stephen J. Gould puts it, "just knowing the properties of each part as a separate entity (and all the laws regulating its form and action as well) won't give you a full explanation of the higher level in terms of these lower-level parts, because in constructing the higher-level item, these parts combine and interact" (Gould, 221). Chaos theory is non-reductionist. James Gleick has written an impressive popular book on the subject. The theory tries to find in our experience of randomness and unpredictability, in a word, the chaotic, a pattern or order that reductionist simplification does not yield. Chaos theory does not abandon determinism or the search for order, nor does it entirely replace the explanatory function of reductionism, but what it tries to do is reconcile our phenomenological experience of the world with the underlying patterns to be found in it. The theory arose out of the sense that certain physical behaviors (weather, for example) could not be explained by reduction to simple elements. Chaos theory tries to preserve our ordinary sensory experience of complexity in daily life at the same time that it discovers an order in it. It is intended as a complement to reductionism, not as substitute for it. I have neither the competence nor the desire to address reductionism as it operates in the natural sciences. My concern is with its application in the humanities.

I want to dissociate my critique of Wilson and his colleagues from that of certain of their adversaries in biology. In *Not in Our Genes: Biology, Ideology and Human Nature*, Richard Lewontin and his collaborators, Steven Rose and Leon J. Kamin (henceforth Lewontin for the sake of convenience) attack sociobiology on grounds that seem more political than scientific. The opposition between Wilson and his adversaries is not a courteous difference between scientists who mutually respect each other's views while differing. It is a difference with political and moral implications in which the specter of racism has been raised. Their target is genetic reductionism in the work of Wilson and his colleagues done before the publication of *Consilience*. Lewontin (henceforth Lewontin for the sake of convenience) is careful to say that the "political philosophy or social position of the exponents of a particular scientific claim is [not] enough to invalidate that claim." He wants only to remind his reader of what philosophers and historians of science like Thomas Kuhn have demonstrated, namely that science does not evolve in a vacuum. Politics, external social and economic history condition its evolution And, of course, science may have political implications, intended or unintended. But Lewontin's polemical strategy belies his cautionary statement, for he begins his attack on his adversaries not by focusing on the internal logic of and evidence for their theories, but by attacking their political motives.

From the very outset, Lewontin connects the advent of sociobiology with the New Right and the celebration of the virtues of competitive capitalism and unfettered individualism, to which he is hostile, thereby prejudicing whatever legitimate scientific objections he may have to the theory. The culprit is genetic determinism:

> The [sociobiologists] argue, for example, that the properties of a protein molecule could be uniquely determined and predicted in terms of the properties of the electrons, protons and other particles, of which it is composed. And they would also argue that the properties of a human society are similarly no more than the sums of the individual behaviors and tendencies of the individual humans of which that society is composed. Societies are "aggressive" because the individuals who compose them are "aggressive," for instance...Ultimately, all human behavior—hence all society—is governed by a chain of determinants that runs from the gene to the individual to the sum of the behaviors of all individuals. The determinists would have it, then, that human nature is fixed by our genes. The good society is either one in accord with a human nature to whose fundamental characteristics of inequality and competitiveness the ideology claims privileged access, or else it is an unattainable utopia because human nature is in unbreakable contradiction with an arbitrary notion of the good derived without reference to the facts of physical nature. (5-6)

What Lewontin and his collaborators overlook in the matter of biological determinism are, on the one hand, scientists of politically conservative persuasion, like the behaviorist B. F. Skinner, who are radically opposed to genetic determinism (indeed, Skinner qualifies as a radical cultural determinist), and, on the other, scientists on the political left, for instance, the linguist Noam Chomsky, for whom language acquisition is a biological universal that has its source in the brain. However one may understand the implications of sociobiology, Lewontin provides no evidence that political ideology *determined* its advent and evolution. For someone whose bête noire is determinism and reductionism, he displays a curious willingness to offer a politically reductionist explanation of the emergence of a theory to which he is opposed.

The more serious charge is that the genetic bias of the theory entails racism. Lewontin notes the fact that the National Front in Britain and the Nouvelle Droite in France (racist movements) cite Wilson, "who claims that territoriality, tribalism and xenophobia are indeed part of the human genetic constitution, having been built into it by millions of years of evolution." To be sure, there may be something in a doctrine that lends itself to corrupt appropriation. But I can imagine Nietzsche contemplating Hitler and Marx contemplating Stalin with incredulity: "This was not we meant at all." Whether or not, Wilson's science is valid on other grounds, there is nothing in his work that implies a *moral* endorsement of territoriality, tribalism, and xenophobia. Wilson would doubtless say that if these are consequences of man's biological life, they have to be dealt with in order to mitigate or overcome them. But there is no evidence that Wilson's work proceeds from racist prejudices, and there is no justification to hold him responsible for the appropriations of his work by the racist movements in Britain and France, just as there would be no justification for holding Nietzsche responsible for Nazism or, more to the point in the case of Lewontin, Marx for all the horrors of Stalinism. In his criticism of cultural relativism, Wilson leaves little doubt about where he stands on matters of race.

> Where cultural relativism had been initiated to negate belief in hereditary behavioral differences among ethnic groups—undeniably an unproven and ideologically dangerous conception—it has then turned against the idea of a unified human nature grounded in heredity. A great conundrum of the human condition was created: If neither culture nor a hereditary human nature, what unites humanity? The question cannot be just left hanging, for if ethical standards are molded by culture, and cultures are endlessly diverse and equivalent, what disqualifies theocracy, for example, or colonialism? Or child labor, torture, and slavery? (201)

Racism is anathema to Wilson's biological universalism.

In the place of sociobiology, Lewontin and his collaborators propose a dialectical model (derived from Marx) in which biology and culture are intertwined, so that neither is reducible to the other. The dialectical model eschews the idea of mere interaction, which assumes the discreteness of biology and culture and "accepts the ontological priority of the individual over the collectivity," though Lewontin acknowledges and praises its anti-reductionism. "Interactionism is [only] the beginning of wisdom" (206). Biology and culture are seamlessly connected in his version of the dialectic. Given the strenuousness with which he *reduces* the work of his biological adversaries to political motives, one may wonder about his own disinterestedness as a scientist in pursuit of the truth. Or is he saying that the political motivation of science is ineluctable and that the laurels go to those whose politics are the most correct and humane? Perhaps nothing better illustrates Lewontin's untrustworthiness as a theorist and critic of theories (I leave out his work as a practicing researcher, which I am told is distinguished) than the following statement: "There is nothing in Marx, Lenin or Mao that is or that can be in contradiction with the particular facts and processes of a particular set of phenomena in the objective world" (quoted in Casti 190). Lewontin is not above gratuitous ad hominem remarks such as: "In reading sociobiology, one has the constant feeling of being a voyeur, peeping into the autobiographical memoirs of its proponents" (206). The distortions and stridency of Lewontin's attack on Wilson and his colleagues diminish whatever persuasiveness his critique might have. My own critique of Wilson and his colleagues, severe as it is, does not challenge his political good faith, but rather the philosophical and practical consequences, already in evidence, of their ambitions for the humanities.

As a humanist, I have to take on trust that reductionism has been a powerful, if not exclusive, method for the making of discoveries in the natural sciences. Its application to the humanities is another matter. In his devotion to the "unity of knowledge," Wilson is driven to incorporate all the disciplines, and it is perhaps to be expected that his own discipline would be the shaping force for the unity. We might pause to consider what is meant by unity and whether it is always possible or even desirable. In his essay "The Disunities of Science," the philosopher Ian Hacking points out that there is no unified view among philosophers and scientists about its meaning. Metaphysical unity is the thought, rooted in the sentiment that the world is one, that all the phenomena that compose it are interconnected. Scientists in their practical work seek out unities of interconnectedness, structure, and taxonomy. The search for a Grand Unified theory is necessarily reductionist, meaning that the sciences are reducible to a master science, "a special type of physics." Quarrels about unity often concern the question of what constitutes a scientific method. "There is no longer consensus on what scientific method is." Many hold that there is no singular method, others hold that method is a myth or an illusion. And then there is the question of the value or disvalue of unity. Hacking has the sensible view that it can be both a good and an evil. The unity or harmony of society may be a good. The unity of the self may or may not be desirable. It can provide a focus for self-expression, but it can also repress desire. The unity of the state can repress freedom and diversity, and the unity of the self may repress desire. What then should we say about the unity of knowledge? If Hacking and others who agree with him are right, if, that is, the unity of the sciences is a dubious proposition, what should we say about the unity of *all* knowledge? If by unity we mean the search for connectedness where it exists, who can quarrel with it? Such a search is at the very heart of the scientific, intellectual, and scholarly enterprises. Poets in their use of metaphor and simile also seek to discover connectedness. But as a totalizing, all-encompassing concept, unity may paradoxically be repressive of the search for knowledge, prematurely closing off avenues of discovery (see Hacking in Galison and Stump, 37-74). In the humanities, achieving unity through reduction may betray the rich complexity of the subjects being studied. Wilson simply assumes what needs to be scrutinized that a totalizing unity through reductionism is an unequivocal good.

What follows is a critical examination of a series of attempts to bring to bear on literature, religion, ethics, and history (provinces of the humanities) mostly, but not exclusively, Darwinian perspectives in the interest of achieving scientific understanding. The results are more often than not scientism rather than science. Scientism, according to Karl Popper is "the imitation of *what certain people mistake* for the method and language of science." He mentions F. R. Hayek's use of the term as "the slavish imitation of the method and language of science" (Popper, 105). Tzvetan Todorov characterizes scientism as a reduction of human thought and action to laws of nature, which results in a loss of freedom (4). My own use of the term refers to the failure of certain scientists and non-scientists to respect the boundaries that divide the natural sciences from the humanities. The benefits of science are immense, and I have no sympathy with the condescension toward science one finds too often among humanists—as in the following remark by the editor of *American Scholar* in his introductory column of the summer 2005 issue: "Science matters, friends, although it pains me to say so." I join with the defenders of science in the current battle against the advocates of creationism and intelligent design, the know nothing (or worse, the self-serving cynical) ignorers of global warming, the benighted opponents of stem cell research. My target is hubris, not the genuine practice of science.

1
Reducing Literature and the Arts

Wilson's comments about the arts are animadversions in a work devoted to many other matters. A recent book David P. Barash and Nanelle R. Barash announcing itself as "A Darwinian Look at Literature" with the title *Madame Bovary's Ovaries* is an ominous foreshadowing of what we may expect in the future. Wilson praises its prose style and characterizes it in a blurb as an "account of an important new development in literary criticism: the incorporation of the biology of human nature." David Barash is a professor of psychology at the University of Washington and his daughter Nanelle is a student of literature and biology at Swarthmore College. Their lack of literary sensibility shows in the witless vulgarity of their slangy prose. "Othello is a story about a jealous guy. Huck Finn is a rebellious headstrong boy. And Madame Bovary is a horny married woman" (1). The choice of style (if indeed it is a choice) is a deliberate affront to literary decorum, which the authors may regard as objectionable only to academic literary stuffed shirts. But the vulgarity of the prose goes deeper than its slanginess. Each of the sentences misrepresents a character. Othello is not by nature jealous; jealously arises only when he is "perplexed in the extreme." He is not a guy or an ordinary Joe, but a heroic general of extraordinary quality and eloquence. "Guy" does not appear in anyone's speech about him. Later on in the Darwinian spirit of the book, Othello is referred to as "a dominant bull elk and silverback male." There is, to be sure, animal imagery in the play: "a black ram is tupping your white ewe," but this is the slandering racist outcry of an abusive unnamed citizen. Huck Finn is not a rebellious headstrong boy. He has his rebellious moments, but he is also convention-bound and conscience-ridden. When he breaks the law and allows Jim to go free, he sees his act in conventional terms as a sign that he is damned. "All right, then, I'll go to hell." Most egregious, however, is the characterization of Emma Bovary. "Horniness" reduces Emma to her sexual appetite and utterly ignores the bookish romantic idiom that nourishes her fantasies. Like Anna Karenina and other romantic women in nineteenth-century fiction, she has dreams of a life beyond the perceived squalor of conventional provincial existence. She is in the line of Don Quixote and other fantasists of fiction. The Barashes seem to have completely missed the significance of the epigraph of their book. *"Madame Bovary, c'est moi."* Flaubert, the author of the novel and the sentence, was neither female nor married.

The Barashes exhibit no interest in the language of a work, the very medium of the literary imagination. What matters is evolutionary theory and presumably how it is illustrated in a work, presumably, because too often the theory has little to do with the text at hand:

> Like the rest of us [Othello] has been culturally restricted to just one Desdemona at a time. It should be clear that it doesn't really matter whether such a male is ostensibly monogamous or polygynous: either way, it pays for males to be sexually jealous and thus highly protective of their reproductive prerogatives. During the rut, bull elk are notoriously aggressive and intolerant of each other, while cow elk are comparatively placid; even among monogamous songbirds, males regularly patrol their territories, alert for intruders....It is also worth noting that Othello is considerably older than Desdemona, a pattern that lends itself to an additional slate of evolutionary insights. Yet another biological asymmetry between sexes is that whereas women go through menopause, men remain potentially reproductive into old age. (20, 21)

Where in the play is there the suggestion that Othello feels restricted in his love for Desdemona, and what do the rutting habits of bull elk have to do with Othello's behavior in the play? We do not need Darwinism to perceive that Othello is older than Desdemona or that women go through menopause while men remain potentially reproductive into old age. Neither Desdemona's menopause nor Othello's putative geriatric infertility figures in the play. The Barashes are simply off and running with their evolutionary gender theory without regard for its relevance. It's as if an unintelligent Iago were reading the play, or, if it weren't for the obvious seriousness of the authors and their supporters, we were being treated to a parody of literary criticism.

The Barashes distinguish their psychological approach from "two equally unhelpful poles: the semi-mystical mythologizing of Freud and the sterile behaviorism of John Watson and B. F. Skinner" (3). As far as I know, there is no behaviorist literary criticism in the Watson/Skinner mode. It is to their credit that they did not extend their domain to works of literature. Freud and the Freudians did—and with some justification. Freud, after all, learned a great deal from the literary classics—that is, from Greek tragedy, Shakespeare, the romantic poets, and Dostoevsky, among others. His own work is a literary achievement and has a place in the literary tradition. It sits there perhaps more comfortably than in science. I am not a particular fan of the Freudian approach to literary works, but when exercised with tact, it yields insights. There is an affinity between Freudian psychology and the psychology one finds in literary works. The same can be said for Marxist criticism when it is not vulgar. Marx was a great admirer of the classics, in particular of Greek tragedy and Shakespeare. He found in Balzac, a political conservative, a confirmation of his own vision of society. Like the Freudian, the Marxian world is a human world, a construction of human beings, which literature represents. The Darwinian literary world as imagined by the Barashes is a jungle or a zoo. The Barashes state that no literary work should be judged "by its consistency with any particular theory, even evolution....Indeed, writing that sets literary achievement second to any other goal is inevitably second rate" (203). This is the sort of perfunctory disclaimer that allows them to proceed in spite of it.

Darwin's view of literature is of some interest. As Gillian Beer has shown, Darwin's rich descriptions of the natural world in *The Voyage of the Beagle* was influenced by the his reading of *Paradise Lost*, a fact not devoid of irony given Milton's Christian vision of the origins of human life. However, as Darwin grew older his taste for the arts (literature and music) atrophied. In a letter he wrote in 1868 and in his autobiography of 1876 he speaks of his soul as having become "too dried up to appreciate Handel's Messiah" (quoted in Donald Fleming, 573). In his later years, the reading of Shakespeare nauseated him. He lamented the fact that his "mind seems to have become a kind of machine for grinding general laws out of large collections of facts." And he attributed the loss of a taste for the arts, "a loss of happiness," to science: "It sometimes makes me hate science." To be sure, Darwin's biography is hardly decisive evidence that Darwin's theory might not be useful in the study of the arts. Elsewhere I caution against the genetic fallacy, the reduction of an idea or an achievement to its origins or motives. Moreover, Darwinism as a cultural force was a significant presence in the literary imagination of the nineteenth century. But it is a matter of interest that Freud and Marx, reductionists each in his own way, were greater appreciators and even interpreters of the arts. Darwin's testimony to the deleterious effects of his scientific obsessions might throw some light on why Darwinism does not appear to be a resource for the understanding of the arts.

One might dismiss the vulgarity of *Madame Bovary's Ovaries* as an aberration, but the support it receives from Wilson and others suggests that this is not the case. And yet it would misrepresent literary Darwinism simply to identify it with its version of literary criticism. In his chapter on the arts in *The Blank Slate*, Steven Pinker brings to bear a conception of *human* nature, presumably derived from evolutionary psychology, on his evaluation of the arts. Here is his summary statement of the project of evolutionary psychology: "It holds out the hope of understanding the *design* or *purpose* of the mind—not in some mystical or teleological sense, but in the sense of the simulacrum of engineering that pervades the natural world. We see these signs of engineering everywhere: in eyes that seem designed to form images, in hearts that seem designed to pump blood, in wings that seem designed to lift birds in flight" (51). In addressing the arts, however, Pinker derives his categories or criteria not from the science of evolutionary psychology, but from the philosopher Denis Dutton, who has "identified seven universal signatures of art."

1. Expertise or virtuosity. Technical skills are cultivated, recognized and admired;
2. Non utilitarian pleasure. People enjoy art for art's sake, and don't demand that it keep them warm and put food on the table.
3. Style. Artistic objects and performances satisfy rules of composition that place them in a recognizable style.
4. Criticism. People make a point of judging, appreciating, and interpreting works of art.
5. Imitation. With a few exceptions like music and abstract painting, works of art simulate experiences of the world.
6. Special focus. Art is set aside from ordinary life and made a dramatic focus of experience.
7. Imagination. Artists and their audiences entertain hypothetical worlds in the theater of the imagination.

These are reasonable and recognizable, if very general statements, to describe "the making and appreciating of art" (Pinker, 404). Unlike the Barashes, Pinker means to occupy a ground familiar to critics of art and literature. His purpose is to judge the arts according to a conception of human nature that has the sanction of evolutionary psychology. Unhappy about the condition of modern art, he wants to answer the question of why "the arts are in trouble." He finds the answer in Virginia Woolf's statement, occasioned by a recollection of "a London exhibition of the paintings of the post-Impressionists, including Cezanne, Gauguin, Picasso and Van Gogh," that human nature had changed in 1910, and, Pinker remarks, "modernism certainly proceeded *as if* human nature had changed" (409-10). But human nature has not changed. In its false conception of human nature, modernism, according to Pinker, has produced artistic travesties. In painting, realism gives way to "freakish distortions of shape and color and then to abstract grids, shapes, dribbles, splashes" and so on. Evolutionary psychology and cognitive science tell us what human nature is, and art required by that nature has given us narratives with beginnings, middles, and ends and by unfreakish representations or imitations of the real world. So much for the work of Cezanne, Picasso, Joyce, Eliot, Proust, and of Kafka, who imagines a human being (unnaturally!) transformed into a gigantic insect. How then should we account for the existence of such art? Pinker views modernist art as a pathology and his diagnosis is "masochism." (Shouldn't it be at least a question whether masochism, even if one were to grant this reading of modernism, is part of human nature and might require an *un*moralizing analysis?)

Pinker wants us to believe that he speaks in the name of evolutionary science, but in fact his guru in art criticism is the sometime-philistine critic Tom Wolfe. In his exposé of contemporary painting, he, like Wolfe, makes no discriminations among the following artists: Pollock, de Kooning, Newman, Noland, Rothko, Judd, Johns, Olitski, Louis, Still, Kline, Frankenthaler, and Stella. They are all "completely literary, the paintings and other works exist only to illustrate the text" (414). Assuming for the sake of argument that Wolfe and Pinker are correct in their characterization of much contemporary painting as literary, why is it a violation of the imperatives of human nature? How does programmatic art of this kind differ from Christian art of the Middle Ages in which paintings existed to illustrate scripture. The fact is that both Wolfe and Pinker are here betraying an anti-modernist bias against abstract art in attacking modernism. Pinker, of course, has a right to his taste. The *representational* and programmatic art of medieval painting may perhaps provide him with a satisfaction not to be found in contemporary painting, but the effect of invoking the authority of human nature, as propounded by evolutionary psychology, is to turn his personal preferences (his taste) into an objective scientific assertion. In which case, there would no longer be a place for alternative tastes and interpretations and evaluations that have been the staple of literary and artistic interpretation.

Pinker does make a half-hearted effort to understand the motives of modernist artists, when he speaks of their attempts to disturb the complacency of the Victorian era the naïve bourgeois belief in certain knowledge, inevitable progress and the justice of the social order. He even concedes that scientific developments may have had something to do with literary and artistic modernism, but he cannot refrain from a condescending tone in characterizing that influence. "*According to the version that trickles into the humanities*, Freud shows that behavior springs from unconscious and irrational impulses, Einstein showed that time and space can be defined only relative to an observer, and Heisenberg showed that the position and momentum of an object were inherently uncertain because they were affected by the act of observation" (my emphasis, 410). Does Pinker mean to deny the rightness or the significance of this version? He doesn't say. In any event, having decided on the unnaturalness of modernist art, he feels no obligation to undertake a serious and disinterested inquiry into its historical *evolution*. He then extends his indictment to postmodernism, which he somewhat ignorantly views as an extension of modernism in its "[more] aggressive relativism" and "its denial of meaning, knowledge, progress and shared cultural values." Somewhat ignorantly, because, though postmodernism continues certain themes of modernism, it is also a reaction against modernist cosmopolitanism and universalism. He betrays a total ignorance of modernism when he speaks of "the affectation of social reform that surrounds modernism and postmodernism" (415). Where, one wonders, is the reformism in the work of Eliot and Joyce? He reaches the heights or depths of absurdity in speaking of postmodernism as "more Marxist and far more paranoid, asserting that claims to truth and progress were tactics of political domination which privileged the interests of straight white males" (411). "More Marxist" incredibly implies the lesser Marxism of modernists such as Eliot, Pound, and Joyce. And how would Pinker reconcile the relativism of postmodernism with Marxism, which affirms the possibility of objective truth and progress? His authority as a linguist and an evolutionary psychologist has apparently exempted him from the obligation to provide a minimally accurate account of the historical phenomena that he is attacking.

One may sympathize (I do) with Pinker's criticism of the deconstructionist view that "there is nothing outside the text" and "that we inhabit a world of images rather than a real world." The denial of reality outside the text is either hyperbolic or false, but must we choose, as Pinker wishes us to, between either-or where images and reality are concerned? Though not identical, they have a relation to each other. There are persons, for example poets, who live in their heads, that is, in a world of images that may be "alienated" from the conventionally received view of the real world. Writers such as Henry James and Vladimir Nabokov often enclose the "real" in quotation marks because they wish to suggest not that there is no such thing as reality, but that the conception of it may be contested—or that one man's reality may be another man's illusion. Postmodern deconstructionists are legitimate targets when they assert the ubiquity of the image world and the unreality of the object world, but an alternative view need not dismiss the reality altering power of images. Shouldn't we want to allow a place for the extravagance of Blake's assertion that the world is not real until it has been transformed by the imagination? Consider its fruitful results in Blake's poetry. The evolutionary psychologist's (or at least this evolutionary psychologist's) conception of human nature, applied to the arts, produces the most reductive, restrictive and formulaic prescriptions and proscriptions. Invention goes by the board, as does Denis Dutton's seventh universal signature: "Imagination. Artists and their audiences entertain hypothetical worlds in the theater of the imagination." Pinker may be right (indeed, I believe he is right) to assert that the mind is not a blank slate, but it does not follow that its inscriptions are unchanging. From what he gives us in his chapter on the arts, I for one don't have much confidence in what he tells us is on that slate.

The effect of Pinker's evolutionary aesthetic on the arts would be conservative, should it ever take effect, since it views traditional forms as conforming to its conception of human nature. Pinker mentions with approval the emergence of a new "movement" in the arts called the "Derriere Guard, which celebrates beauty, technique and narrative," and goes on to cite other movements, "The Radical Center, Natural Classicism, the New Formalism, the New Narrativism, Stuckism, the Return of Beauty, and No Mo Po Mo" (417). Since we are given no specific names of artists and their achievements, we have no way of judging the value of these movements. It is something of a paradox that an evolutionary psychology should have conservative consequences. To be sure, evolutionary change is slow moving and not radical, but doesn't evolution imply that human nature may not be fixed (not the same thing as saying that it is blank slate)? If an artist (and indeed more than one artist) feels compelled to undermine representation and narrative, can't that compulsion or impulse be seen as an expression of the changing cultural nature of humanity or a portion of it? What is unnatural about a painting illustrating a theory or a text, to take the instance provided by Tom Wolfe's critique of modern painting? One may have all kinds of legitimate objections to the particular performances of the painters he cites, but not on the grounds that the painting is unnatural. It is in the nature of human beings or, to put it more cautiously, certain human beings, to transgress norms. One suspects that Pinker is confusing or conflating norm and nature. To do so is to engage in a dangerous activity, for we know that norms can become oppressive and that they can change. (It was not very long ago that homosexuality was universally regarded as unnatural and deviant.) Human nature may not be a blank slate, but do we know enough to know what is inscribed upon it? In and of itself a conservative view of the arts may have much to recommend it, but only if it proceeds from aesthetic and intellectual judgment and not from an a priori conception of human nature or, from what may be the case, socio-cultural prejudices disguised as science. We simply do not have a sufficient knowledge of human nature to prescribe or proscribe art on the basis of it. Indeed, without prescription of proscription, we expect writers and artists to explore the uncharted territory of human nature.

In reading the following passage in *The Blank Slate*, think of Stendhal or Flaubert or Baudelaire:

> As for sneering at the bourgeoisie, it is a sophomoric grab at status with no claim to moral or political virtue. The fact is that the values of the middle class—personal responsibility, devotion to family and neighborhood, avoidance of macho violence, respect for liberal democracy—are good things, not bad things. Most of the world wants to join the bourgeoisie, and most artists are members in good standing who adopted a few bohemian affectations. Given the history of the twentieth century, the reluctance of the bourgeoisie to join mass utopian uprisings can hardly be held against them. And if they want to hang a painting of a red barn or a weeping clown above their couch, it's none of our damn business. (416)

So much for a disinterested scientific approach to the subject. Pinker defends the bourgeois virtues, which are real, but he writes about them as if they are pieties immune to criticism. If Pinker had read or reread Flaubert or Baudelaire, he would have realized that you need not be utopian to be anti-bourgeois, or that you can be bourgeois and anti-bourgeois at the same time. He is disturbed by the "[im]moral and political track of modernist artists," as he has a right to be, but he has no sense of how an artist can be something different in his art from what he is in his life. The ethical judgments that we make about the conduct of an author's life are not immediately transferable to his art. Perhaps the doctrine of the impersonality or autonomy of art is beyond the understanding of evolutionary psychology. Imagine a cultural ethos that proscribes a literature that disturbs bourgeois complacency and the criticism of bad taste. "None of our damn business" belongs in the barroom. (And yet Pinker makes it his business to castigate the taste for modernism, indeed to view it as a pathology. Should we say "none of his damn business"?) It would be a cultural misfortune if Pinker's prophecy came true: "I predict that the application of cognitive science and evolutionary psychology to the arts will become a growth area in criticism and scholarship" (418).

His usual confidence that evolutionary psychology possesses the truth about human nature is occasionally qualified by an admission not only that there is much that is not yet known about the mind, but also that "our world might always contain a wisp of mystery, and our descendants might endlessly ponder the age old conundrums of religion and philosophy, which ultimately hinge on the concepts of matter and mind" (240). He concludes the chapter, "Out of Our Depths," with a citation from Ambrose Bierce's *The Devil's Dictionary*: "**Mind**, *n*. A mysterious form of matter. Its chief activity consists in the endeavor to ascertain its own nature, the futility of the attempt being due to the fact that it has nothing but itself to know itself with." Bierce is not an advocate of the blank slate. His challenge is more profound. *The nature of mind is and will probably remain a mystery.* Pinker concedes to it by saying that "the world might always contain a wisp of mystery." Why only a wisp?

Neither Pinker nor the Barashes are literary scholars; they are interlopers from evolutionary psychology. Nor, as we have seen, do they approach the subject of literature in the same way. What they do have in common is an impoverished sense of the literary imagination. We should expect greater sensitivity to the variety of literary expression from literary critics and scholars who have adopted the Darwinian approach. In her essay "Making Knowledge: Bioepistemology and the Foundations of Literary Theory," the literary scholar Nancy Easterlin proposes, as her title promises, bioepistemology as a theory to take the empty place created by the waning of postmodernism in the academy. She is one of a growing number of literary scholars who have adopted the cause of neo-Darwinism in literary study, among them Joseph Carroll, the author of *Evolution and Literary Theory* and Robert Storey, author of *Mimesis and the Human Animal: On the Biogenetic Foundations of Literary Representation*. What that theory amounts to in Easterlin's formulation of it is a recognition of the adaptive function of narrative-making in organizing and controlling reality: "If narrative principally serves to give coherent shape to the events of social life, the imputation of necessity to nature suggests a profound disposition to discover causal order in nature. Both are adaptive tactics, because the intelligibility of events resulting from narrative construal is correlated with a feeling of control and mastery"

(Patai and Corral, 628). It is not as if we are learning something new in this recognition. Writers and critics have long spoken about the therapeutic (i.e., adaptive) function of literature. So the confidence of the new breed of bioepistemologists that they will revolutionize our knowledge of every discipline is hard to fathom.

> "Positive scientific understanding," Carroll insists, will come to dominate even the humanities, and it will do so, "in spite of all prejudice and all entrenched interests," because "of the irresistible force of its explanatory power." Carroll is alluding to a specific kind of understanding—to the Darwinian paradigm that informs this book—and his argument is precisely my own that "the evolutionary explanation of human experience is...a more complete and adequate theory of the development and nature of life, including human life, than any other theory currently available to it," and that it "thus necessarily provides the basis for any adequate account of culture and of literature." Where its eventual triumph will leave poststructuralism and its disciples is, intellectually and professionally nowhere, for the poststructuralist paradigm "operates on principles that are radically incompatible with those of evolutionary theory" (Storey, 205-06).

As for "traditional humanist criticism," Carroll is dismissive of its dualistic separation of "the physical world and subjective experience." He finds it wanting in its unsystematic impressionism. He cites Roger Shattuck's "traditional" humanist view that "the arts and the humanities do not look primarily for universals and general laws: they seek out the revelation and uniqueness of individual cases" (quoted in *Literary Darwinism*, 77). A careless reader, Carroll concludes that "if we were to take the proposition at face value, we should be unable to make any generalizations about or comparisons of literature. We should be unable to identify commonplaces of period, style, or of genre." On the contrary, Shattuck does not say that the arts and the humanities do not make generalizations, he does say that the *primary* task is not to look for "universals and general laws." Nothing in "traditional humanist criticism" proscribes generalizations about period, style, or genre. But these generalizations do not rise to the condition of laws. "Traditional humanist criticism" is an unfortunate coverall for a diversity of critical approaches that includes John Dryden, Dr. Johnson, Samuel Taylor Coleridge, Matthew Arnold, I. A. Richards, William Empson, Lionel Trilling, F. R. Leavis, John Crowe Ransom, Raymond Williams, et al. It is dismaying to have to encounter in a literary scholar, who should know better, a dismissive attitude toward what has already been accomplished in the humanities. And on other occasions, he does know better. An author of books on Matthew Arnold and Wallace Stevens, he is unhappy when Steven Pinker characterizes the pleasures of literature as "mental cheesecake." When he turns to a work of literature, for instance *Pride and Prejudice*, he has sensible, though not particularly original, things to say, which, one should note, do not depend on an application of evolutionary theory. "Austen mocks false status—rank and wealth unsupported by education, wit, manners and character—but she ultimately affirms the authority of legitimate social status as that represented by the normative couple, Elizabeth and Darcy" (213). Nor do we need evolutionary theory when he tells us that "irony is a fundamental and pervasive literary device designed for the purpose of detecting and exposing hypocrisy and deceit" (81-82). It needs to be said that very little of the vitality and wit of *Pride and Prejudice* gets into Carroll's account of the novel. Carroll's complaint about traditional criticism, "which operates at the level of Austen's own lexicon… is that it seeks no systematic reduction to simple principles that have large general validity" (213). This hardly seems a deficiency. The alternative that Carroll and his fellow literary Darwinists are proposing is the dissolution of the individuality of a work (the very reason we enjoy and value it) into a large generalization that removes all of its distinctive features and vitality.

Literary Darwinism has already entered the mainstream media. The Sunday magazine section of the *New York Times* (November 8, 2005) published an article by D. T. Max titled "Literary Darwinists: Can Evolutionary Principles Shed New Light on the Literary Canon?" On the basis of interviews with some of the practitioners of this new approach, Max's answer is a qualified yes. Here is his summary of the program: "Literary Darwinists read books in search of innate patterns of human behavior: child bearing and rearing, efforts to acquire resources (money, property, influence) and competition and cooperation within families and communities." They look for "certain universal ways and do so because those behaviors are hard-wired into us" (76). Max offers as an example of a universal, the celebrated opening sentence of Jane Austen's *Pride and Prejudice*: "It is a truth universally acknowledged that a young man in possession of a fortune must be in search of a wife." Of course, as Max and the literary Darwinists whom he has interviewed know, the sentence is ironic. (Incidentally, can evolutionary theory account for the emergence of irony?) The truth, according to the novel, is that it's "the women who aggressively compete to marry high status men, consistent with the Darwinian idea that females try to find mates whose status will assure the success of their offspring." How universal is *this* truth? Does it apply to all females and to all historical periods? Evolutionary theory is presumably a theory of history, but its historical perspective is suspended when it focuses on the human condition. *Pride and Prejudice* deals with a particular period and a particular section of English society. The unaddressed question is how the presumed innate patterns of human behavior are inflected by particular social conditions? Elizabeth Bennett and her friend Charlotte Lucas have uppermost in their minds the fear of homelessness if they are not married. And that fear has its source in a historically conditioned socio-economic fact. The Bennett estate is entailed to a male cousin; the fear is that once Elizabeth's parents die she will be forced to leave her home and find employment as a governess. Charlotte, who is in a similar position, settles for Mr. Collins, the foolish cousin on whom the Bennett estate is entailed. In contrast, Elizabeth's refusal to marry Mr. Collins, who had earlier proposed to her, reflects a sense of self and of the worth of marriage that trumps her fear of homelessness. What we have then are women of comparable intelligence with different responses to the same situation. They cannot be reduced to a Darwinian hard-wired universal pattern.

Perhaps literature may provide information for biological understanding, as Jonathan Gottschalk, editor of the anthology *The Literary Animal*, implies: "The thing literature offers is data. Fast, inexhaustible, cross-cultural and cheap" (Max, 76). But, as the very language of this claim should make clear, literary Darwinism does nothing to illuminate our literary understanding or our understanding of historical context. Max, the author of the *Times* article, apparently thinks otherwise, despite his acknowledgement of its inadequacies. "Ultimately literary Darwinism may teach us less about individual books than about the point of literature" (78). Needless to say, literature, like life itself, does not have a point, which is not to say that it is pointless.

E. O. Wilson writes: "The love of complexity without *reductionism* makes art, the love of complexity with *reductionism* makes science" (59). But the wisdom of this statement fails to carry over to the program he proposes for humanistic and literary study, as should be evident from the practices I have just described. In Wilson's proposal for the study of literature its complexity would simply evaporate. Let's assume that natural selection, the driving force of evolution, explains the emergence of a brain that has the capacity to produce thought and art. What it can't account for is the autotelic, non-functional, non-adaptive character of thought and art. Indeed, the making of art is in great part a product of intelligent design, not of natural selection. As the literary scholar Dwight Culler pointed out years ago, "the Darwinian-Benthamite-Malthusian view" is "antithetic to the purposive cast of the human mind" (597).

As I have already remarked, it should not surprise us that the humanities may be vulnerable to another theoretical takeover. Evolutionary theory is only the latest of a series of theories that have invaded or tried to invade the humanities in recent decades, though it may be the first to have or presume to have the authority of the natural sciences. Northrup Frye in *An Anatomy of Criticism* in fact analogized his work to that of evolutionary biology, but he did not write with the authority of a natural scientist. What he had in mind was the achieving of a discipline comparable in its coherence to a science. Though he failed to realize his ambition, he had the advantage of being a literary scholar of great distinction, so his work is a remarkable achievement full of literary insight. If literary study is vulnerable to what I call theoretical takeovers, it also tends paradoxically to be resistant to theory, because it has always affirmed the individual, not the generic response to a work. The prevailing pronoun of theoretical response is "we," not "I." The humanities may suffer from an inferiority complex vis-à-vis science in its cultivation of rigor and in the tangibility of its discoveries and applications, but the humanities are also possessed of the conceit (meant unpejoratively here) that its domain is that of imagination, feeling, irreducible complexity, and inwardness that defies the reductions of theory. Some of our most distinguished literary critics have resisted the temptations of theory in the interests of individual intuition and insight. What counts for them is the quality of critical intelligence and literary sensibility. It is not that there is no community of literary insight, but that each reader or critic is encouraged to find the differences within a work that reflects his or her own cast of mind and experience. T. S. Eliot said that the one thing needful for a literary critic is intelligence, and F. R. Leavis agreed. Just as there is no theory to encompass all poems and/or novels, there is no theory to *govern* the responses of readers and critics. Denis Donoghue insists on speaking of poems, not poetry and Christopher Ricks on affirming the role of principles, not theory. Like Shattuck, they affirm the uniqueness of the individual work,

If the presence of theory in the humanities has been a source of a certain discomfort, it is nevertheless inescapable in any discipline. What are its powers and limitations? As Geoffrey Hartman has justly remarked, literary theory is canon-specific. Beware of the grand theory that tries to encompass the totality of literature and in consequence erase the differences that manifest its vitality. Frye's structuralist *Anatomy*, for all its explanatory achievements, is necessarily reductive. Which is not an argument against theory per se, but rather a statement about its limits.

No theory depends exclusively on empirical induction. Every theory has its share of a priorism. The positivistic belief that theory should have its exclusive origins in (and dependence on) empiricism is an illusion, but at the same time theoretical speculation should find anchors in the actual experience of literature. Generalizations about genre, narrative, image making, and reader response are inevitable and desirable, but they must at once be modest, relevant, and constructive—and not inhibit and impoverish the individual responses of readers and spectators. Literary Darwinian lacks these requisite qualities.

Postscript: I have confined myself to Darwinian hubris in literary criticism. Occasionally, an exponent of evolutionary theory and cognitive science will claim for it the potential power of poetic imagination. In *Darwin's Dangerous Idea*, Daniel Dennett speaks of the evolutionary process as algorithmic, which means one of two things: (1) a logical step-by-step procedure for solving a mathematical problem in a finite number of steps, often involving repetition of the same basic operation; (2) a logical sequence of steps solving a problem written out on a flow chart, that can be translated into a computer program. The algorithm is a feature of Artificial Intelligence, of which Dennett is an advocate, as well as he is of evolutionary theory. His blithe confidence in what the algorithm can achieve is breathtaking. "If the algorithmic process is powerful enough to design a nightingale and a tree, should it be that much harder for algorithmic process to write an ode to a nightingale or a poem as lovely as a tree" (451). Well, the algorithm could conceivably produce a sequence of words that some might want to call a poem. But it is extremely doubtful, if not incredible, that it could produce Keats's poem or, for that matter, even Kilmer's, which is not as lovely as a tree. Any poet, worthy of the name, will tell you that the making of a real poem is not a logical sequence of steps. Moreover, the algorithm gives us no information about the difference between a good poem and a poor one. As we might expect, Dennett is uncomfortable with settling for "intuition" as an explanation of how the mind in one of its moods works. "Whenever we say we solved some problem by 'intuition' all that really means is *we don't know* how we solved it" (442). He wants to reduce intuition to a yet undiscovered algorithm. The effect would be to evaporate the very idea of what Michael Polanyi has called tacit knowledge, which is present in all disciplines. Should we want that to occur? T. S. Eliot did not think so when he meditated on the creative process in the making of poems. He said that the bad poet is conscious of what he is doing when he should be unconscious and unconscious when he should be conscious. (I am assuming that intuition has its source in the unconscious part of our being.) Could it be that intuition is protean and that no algorithm could subsume its variety? Roger Penrose, the mathematical physicist, is a skeptic: "I am a strong believer in the power of natural selection. But I do not see how natural selection, in itself, can evolve algorithms which could have the kind of conscious judgments of the validity of other algorithms that we seem to have" (Dennett, 444). Our capacity to create, interpret, understand, and evaluate artistic and intellectual achievement springs from our imaginative and intellectual faculties, which may have their origins in natural selection, but whose functioning and achievements cannot be explained by it.

2
Demystifying Religion

Nothing in our cultural life these days is more fraught with anxiety than the relations between science and religion. "Relations" is a euphemism for the boundary-eroding antagonism that exists between them. In its fundamentalist incarnation, religion wants to substitute creationism for evolution; in its rationalist militancy, science dreams of the withering away of religion and its illusions. If the main burden of responsibility for the present state of affairs lies with the religious fundamentalists, the militant atheism of certain prominent exponents of the scientific worldview has only exacerbated matters. In *Imagined Communities*, Benedict Anderson, a historian with Enlightenment and Marxist sympathies, turns out to be an eloquent witness for the religious imagination against its detractors in the camp of scientific progress:

> The great merit of traditional religious worldviews (which naturally must be distinguished from their role in the legitimation of specific systems of domination and exploitation) has been the concern with man-in-the-cosmos, man as species being, and the contingency of life. The extraordinary survival over thousands of years of Buddhism, Christianity or Islam in dozens of different formations attests to their imaginative response to the overwhelming burden of suffering—disease, mutilation, grief, age, and death. Why was I born blind? Why is my best friend paralyzed? Why is my daughter retarded? The religions attempt to explain. The great weakness of all evolutionary/progressive styles of thought, not excluding Marxism, is that such questions are answered with impatient silence. At the same time, in different ways, religious thought also responds to obscure intimations of immortality, generally by transforming fatality into continuity (karma, original sin, etc.). In this way, it concerns itself with the links between the dead and the yet unborn, the mystery of re-generation. Who experiences *their* child's conception and birth without apprehending a combined connectedness, fortuity, and fatality in a language of "continuity"? (Again, the disadvantage of evolutionary/progressive thought is an almost Heraclitean hostility to any idea of continuity.) (18-19)

The evolutionary/progressive style of thought turns away in silence from the heartrending questions in Anderson's list, because there is nothing in progressive or scientific thought that can provide an answer to these questions. They are not questions for science, because by their very nature there are no scientific answers to them. The religious "attempt to explain" when not performed in a dogmatic style may be necessary and valuable for a large portion, perhaps even a majority, of humanity who seek not objective truth in the answers, but consolation. Anderson here is not taking sides in the culture war between religion and science. There is a long and awful history of suffering brought about by religions packed into the phrase "their role in the legitimation of specific systems of domination and exploitation." Anderson's own work is in the progressive style of thought, but it is a style modified by an awareness that it too can legitimate "specific systems of domination and exploitation," if it is not aware of its own limitations. The passage suggests the need for complementary perspectives.

In *Man Stands Alone*, the evolutionary biologist Julian Huxley places religion within the evolutionary process:

> In one period of our Western civilization the gods were necessary fictions, useful hypotheses by which to live.... The disasters of the outer world must still be sufficiently uncomprehended and uncontrolled to be mysteriously alarming. Or else the beastliness and hopelessness of common life must be such as to preclude any pinning of faith to the improvement of the world; then God can, and social life cannot, provide the necessary escape-mechanism. (277)

Huxley goes on to express a faith, unsubstantiated by science, that the "advances of natural science, logic and psychology have brought us to a stage at which God is no longer a useful hypothesis." By "no longer useful" Huxley presumes an enormous as yet unrealized potential in the progress of science not only to illuminate what has been dark and uncomprehended in the past, but also to overcome the "beastliness and hopelessness of common life." Who would deny the contribution that science has made to "the improvement of the world"? But we have seen too much cruelty and suffering in our time (including the suffering brought about by technological advances) to dispose so easily of other ways of viewing the world. The claim that the world was created by God in six days is not a useful hypothesis, but religious thinking and experience are not exhausted by it.

For all its dismissiveness, Huxley's quasi-anthropological view of religion is civility itself compared to the view of Richard Dawkins, militant atheist, influential exponent of Darwinism, and Britain's current leading public intellectual. He assumes the role of medical diagnostician in characterizing "God [as] a delusion" and "religion a virus." In an interview with Dawkins on April 28, 2005, the interviewer introduces him as the religious right's Public Enemy No. 1. But it would be more accurate to speak of him simply as the enemy of religion. The interviewer tells of a British political debate in which Dawkins appeared on the opposite side of a religious minister. "When the minister put out his hand, Dawkins kept his hands at his side and said: 'You, sir, are an ignorant bigot.'" Perhaps the minister was a bigot, but the interviewer, sympathetic to Dawkins, does not provide us with an account of the minister's views. At the very least, he displayed a courtesy completely absent in Dawkins. The uninformed reader wonders who the intolerant person is. In the interview, Dawkins speaks of the "bad religion" of those opposed to "the idea of evolution" which he contrasts with the beliefs of "sophisticated educated theologians." But the distinction is abandoned when he attacks religion in its totality and the belief in God. He attributes the persistent belief in God to "the child mind," which "for very good Darwinian reasons [is] susceptible to computer viruses… that say "spread me, copy me, pass me on." Since the "the child brain is preprogrammed by natural selection to obey and believe what parents and other adults tell it," it is as susceptible to bad ideas as to good ideas. The "bad ideas, useless ideas, waste of time ideas, are rain dances and other religious customs….The child brain is very susceptible to this kind of infection. And it also spreads sideways by cross infection when a charismatic preacher goes around infecting new minds previously uninfected." Where is the science in this diagnosis of religion as an infection? Was Milton suffering from a disease when he produced *Paradise Lost*? Dawkins here is employing a diagnostic metaphor as an expression of sheer ideological animus.

He tells us that, as an antidote to the childish brain of his young daughter, he has advised her not to believe anything for which no evidence is provided. It sounds reasonable, but highly impractical. In life we learn to trust certain people and mistrust others. Without that trust and mistrust we cannot act in the world. Evidence is not always forthcoming and often we need to guess our way—as reasonably as possible. We non-scientists believe what science and scientists tell us not because we are able to understand the evidence for their claims, but because we have been taught to trust the authority of science. That trust has been in great part justified by the evidence of medical advances and technological achievements about which we see, read, hear, and experience. But there is still a residue of trust (faith, if you like) that is not exhausted by the evidence. No layman could possibly know, even from the comprehensible evidence available to him, that Einstein is right in his theory of relativity or that the fossil record establishes the truth of natural selection. I don't mean to place the layman's trust in the achievements of science on the same level with faith in a divine being. I want only to draw attention to a mindset that is inimical to a faculty that we all share, whatever differences we may have on a host of matters. Dawkins's hostility to all forms of religious expression proceeds from his fetish of always insisting on evidence. (Not that he is always respectful of the evidence in fields about which he is not adequately informed.) He is the Sergeant Friday of Darwinism: the facts, only the facts—as if only the facts are required to arrive at the truth. His is a mindset that prevents him from entertaining the claims of mindsets opposed to his own. It is not enough for him to disagree with his opponent, he wants to annihilate him intellectually.

Lack of respect for evidence is not the only target of Dawkins attack upon religion, it has also been a source of violence and cruelty ("domination" and "exploitation," in Benedict Anderson's words.) And what rational person would deny this? But Dawkins's blinkered vision of the world prevents him from reflecting on the havoc *militant* atheism has wrought in the most terrible century human history has known, wherever it has achieved power. If Dawkins were consistent in his condemnation of violence and unreason of religion, he would also have to face the reality of homicidal tyrants who have acted in behalf of militant atheism, tyrants such as Hitler, Stalin, and Mao. In *The God Delusion*, Dawkins does consider the murderous behavior of Hitler and Stalin, but simply denies that it has anything to do with atheism. (While acknowledging that Stalin was an atheist, he has doubts about Hitler.) It is of course true that atheists by virtue of their atheism don't automatically become murderers just as it is true that theists by virtue of their beliefs do not turn into killers. The problem lies with the fanaticism and militancy with which a doctrine is embraced. Dawkins simply ignores the persecution of religious groups by militant atheists in power, perhaps because his own militancy prevents him from acknowledging it (see 272-8). Dawkins cites with approval Sean O'Casey's preposterous remark: "Politics has slain its thousands, but religion has slain its tens of thousands" (235). Anyone alive in the twentieth century should know that politics has in fact slain millions. It could be argued that certain versions of atheism (not all atheism, to be sure) allow for the incarnation of deity in a tyrant. What else is the political cult of personality but an atheistic religion? And here we may have an explanation of a possible benign effect of theistic religious belief. As I have pointed out in *The Reign of Ideology*, "for East European writers such as Leszek Kolakowski and Czesław Miłosz, who have known Caesar, Christianity has proved to be a recourse from political despotism. In its theocratic form Christianity is itself a despotism, but it may also function as a dissidence, a vehicle of freedom in a secular authoritarian or totalitarian society" (37). I don't mean to suggest an association between all forms of atheism and that of the secular totalitarian societies of the twentieth-century. Indeed, in conflating all atheisms in that manner, I would be following Dawkins in his homogenization of religious beliefs.

Despite his vehemence toward institutional religion and the belief in God, Dawkins cannot entirely dispense with religious sentiment. He acknowledges its persistence in himself when he speaks of a "kind of quasi-religious feeling" for the natural world. In *Unweaving the Rainbow*, he "specifically attacks the idea that a materialist, mechanist, naturalistic world view makes life seem meaningless. Quite the contrary, the scientific worldview is a poetic world view, it is almost a transcendental world view" ("The Atheist," interview with Gordy Slack, 6). Well, this is assertion, not argument, let alone scientific demonstration. What exactly is an "almost transcendental" worldview? The following sentence captures Dawkins's sentiment: "The truths of evolution, along with many other scientific truths, are so engrossingly fascinating and beautiful; how truly tragic to die having missed out on all that" (*The God Delusion*, 238). This is not romantic nature worship. It is evolution that is the object of reverence, a process that includes "nature red in tooth and claw." This from a man who devotes many pages in his book to excoriating the cruelties depicted in the Bible. Or is the object of reverence the laws of nature rather than the activity they describe? In which case, it is an exhaustible reverence, for as Dawkins remarks, "most scientists are bored with what they have already discovered. It is ignorance that drives them on" (125). Of course, he and his fellow scientists should be allowed their faith or reverence, but what if others cannot find spiritual fulfillment in this vision of the natural world? Do they deserve to be treated as pathological cases? The metaphor of pathology for religion is itself like a virus (two can play the game) in neo-Darwinian discourse about religion. Dawkins is not alone in his views. Here is Darwin's fellow biologist George Williams: "There is no more reason to expect a cultural practice transmitted between churchgoers to increase churchgoers' fitness than there is to expect a similarly transmitted flu to increase fitness?" (quoted in Dennett, *Darwin's Dangerous Idea*, 361). "Survival of the fittest" is the test, and religion, according to Williams, fails it. Is ther nothing more to life than sheer survival? Dawkins and Williams are in a tradition that includes Marx, who spoke of religion as an opiate and Freud for whom religion was an illusion, though Freud at least took the "illusion" seriously enough to learn from it. (See Moses and Monotheism).

The case of Dawkins is particularly egregious. If his view of the religious imagination were taken seriously, its impact on our intellectual life would be disastrous. It is not enough for him to dismiss literalist readings of the Bible, he has no patience for "sophisticated theologians who do not literally believe in the Virgin Birth, Miracles, the Transubstantiation or the Easter Resurrection," but who "are nevertheless fond of dreaming up what these events might symbolically *mean*." He compares such "dreaming" to hypothetical scientists who would "desperately" search for a "symbolic meaning" of the double helix, should it prove to be wrong, which would "transcend mere factual refutation" (*Unweaving*, 183). Dawkins loves to characterize the errors and misconceptions of his adversaries as preposterous or absurd. The temptation is strong to apply these adjectives to this comparison (and add the adjective "pernicious"). The Bible has been fruitfully read as literature for more than a century and a half, the legacy of the higher criticism. This would not be possible, if it could not be viewed as myth (with its stock of symbol and metaphor) that reflects the reality of human experience. Dawkins's dismissal of symbolic readings of Biblical events could apply as well to non-Biblical secular literature that refers to Biblical event. He might demur (I am only guessing), saying that fiction and poetry that represent Moses or Christ or the Easter Resurrection do not pretend to be fact. But that is not a sufficient answer, because great works of literature (e.g., *The Divine Comedy*, *Paradise Lost*) treat Biblical literature as if it were fact. Should that disqualify the poems? Without the possibility of a symbolic reading of the Easter Resurrection, we would not have D. H. Lawrence's imaginative appropriation of it in his short novel, *The Man Who Died*, and indeed countless other imaginative (and symbolic) recreations of the event. Without symbolic interpretation, the anthropological approach of Edmund Leach and Mary Douglas to scripture would not be possible. In comparing symbolic interpretation to the putative refusal of scientist to accept "factual refutation" for what it is, Dawkins is mixing apples and oranges. The only explanation for the intellectual irrelevance and crudity of this comparison is that he has allowed his fierce hatred of all forms of religious expression to overwhelm his intelligence. For someone eager to rebuke the ignorance of those who speak and write about science without a genuine knowledge of science, he is strangely unconcerned about sounding off on subjects about which he has apparently spent little time to understand. He comes off as an inverted fundamentalist, a counterpart to the Christian fundamentalist, in his hostility to symbolic interpretation. This is not a new charge against Dawkins and, as one would expect, he rejects it. As he understands fundamentalism, it applies only to those who have a literal belief in scriptural falsehood. But he misses the point. He is at one with the literalist readers of the Bible in insisting that the only way of reading the Bible is to read it literally. Where the religious fundamentalist finds truth, Dawkins finds falsehood. Neither Dawkins nor the religious fundamentalist has an appreciation of the symbolic reading of the Bible. Unlike both evangelical Christians and Richard Dawkins (strange bedfellows in this respect), liberal theologians, literary critics, and anthropologists do not need to have a definitive answer to the question of whether the events in the Bible are literally true. They assume that they may or may not have historically occurred, but are mainly concerned with what they signify in our human experience.

If taken seriously, Dawkins's view would deprive not only liberal theology, but poetry, literary criticism, and anthropology of their legitimacy as interpretative disciplines. (As far as theology is concerned, *tant mieux*, Dawkins would say, for "theology… is not a subject at all" [*The God Delusion*, 57]). Lest we have any doubts in the matter, we need only turn to the opening page of *The Selfish Gene*, where he widens his net to include the humanities: "We no longer have to resort to superstition when faced with the deep problems: Is there meaning to life? What are we for? What is man? After posing the last of these questions, the eminent zoologist G. G. Simpson put it thus: 'The point I want to make now is that all attempts to answer that question before 1859 are worthless and that we will be better off if we ignore them completely'" (1). In the endnote to the paperback edition, Dawkins responds to those who have taken offense at the quotation from Simpson by rubbing it in: "I agree that, when you first read it, it sounds terribly philistine and gauche and intolerant, a bit like Henry Ford's 'History is more or less bunk'. But religious answers apart (I am familiar with them; save your stamp), when you are actually challenged to think of pre-Darwinian answers to the questions 'What is man?' 'Is there meaning to life?' 'What are we for?' can you as a matter of fact think of any that are not now worthless except for their (considerable) historic interest? There is such a thing as just being plain wrong, and that is what, before 1859, all answers to these questions were." If Dawkins and Simpson had contented themselves with the claim that Darwinism has made worthless other answers to the questions, where do we come from? and how have we evolved?, they would have given offense only to creationists. But questions about meaning and purpose are of another order and continue to be the legitimate concern of literature, philosophy, and religion. They are not simply reducible to knowledge about our genetic structure. Could our knowledge of evolution and genetic structure affect our understanding of human meaning and purpose? Nothing in the presentations of Dawkins and his fellow evolutionary theorists persuades me that it would, though it would be presumptuous to rule out the possibility. But would such knowledge make the speculations of Plato, Aristotle, Augustine, Montaigne, Rousseau, as well as poets and novelists worthless? The question doesn't deserve an answer.

Dawkins speaks of his love of poetry, though he spends too much time pointing out errors or prejudices about science in poets and poems. He attaches poetry to the scientific project, acknowledging as he does the inescapability of metaphor in his own exposition of scientific ideas. "Selfish" (as in *The Selfish Gene*) is, after all, metaphoric (as well as anthropomorphic) in modifying *gene*. Metaphor, as Dawkins himself points out, may help in communicating difficult ideas to the uninitiated, and they provide imaginative scaffolding for scientists as they think their way in the formulation of theories. They may also prove unreliable and misleading in what they communicate because of their allusiveness and connotative richness. He "solves" the problem by attributing "bad poetic science" to his adversaries and "good poetic science" to himself and his allies, but without providing grounds for the distinction.

Toward the end of *The God Delusion*, Dawkins devotes a section to "Religious education as a part of literary culture." The motive of his proposal is the "biblical ignorance commonly displayed by people educated in more recent decades than I was" (341). He then proceeds to list a number of phrases and sentences that have their source in the Bible that educated people should know, among them "Be fruitful and multiply," "East of Eden," "Adam's Rib," "The mark of Cain": the list is very long. Why it is important to know these phrases, given Dawkins's utter contempt for the moral and spiritual claims of the Bible, is by no means clear. In the section titled "The 'Good' Book and the Moral *Zeitgeist*" Dawkins attempts to show that virtually every episode in the Old Testament is a story of human depravity and that its heroes, Abraham and Moses, among others, are moral monsters in their unquestioning faith in a ruthless God. Given the unrelieved portrait that we are given of the dark side of the religious imagination, the paragraph that concludes the section on religious education seems little more than cant.

> I have probably said enough to convince at least my older readers [surely, Mr. Dawkins, you must be joking] that an atheistic worldview provides no justification for cutting the Bible, and other sacred books, out of our education. And of course we can retain a sentimental loyalty to the cultural and literary traditions of, say, Judaism, Anglicanism, or Islam, and even participate in religious rituals such as marriages

and funerals, without buying into supernatural beliefs that historically went along with those traditions. We can give up belief in God while not losing touch with a treasured heritage (344).

Nowhere in the book are we given the slightest inkling of what makes the heritage treasured in Dawkins's eyes.

His influence should not be underestimated. We find it in writers whose views in other respects are characteristically moderate. Susan Haack, author of *Defending Science—Within Reason; Between Scientism and Cynicism*, seems to suspend her sense of the limitations as well as the powers of science in addressing the theme of religion. For the most part, she is admirably poised between scientism and the new cynicism, a term that covers postmodern challenges to scientific objectivity. But on the question of the rival merits of science and religion, she finds herself in Dawkins's camp. In concluding her discussion of science and religion. she anticipates and rejects the charge of scientism, but doesn't successfully refute it. She quotes with approval Dawkins's view that the appeal of religion is to what is "childlike" in us. "As Dawkins points out, for good biological reasons small children are necessarily credulous—they are 'information caterpillars' sucking in information as caterpillars suck in cabbage leaves" (293). Evidently these good biological reasons do not or should not (!) apply to adults. But what if masses of adult human beings persist in their credulity, shouldn't a biologist try to understand the phenomenon without moralizing about its human value? Haack acknowledges that "religion is no less quintessentially human as an enterprise than science," but attributes "its fundamental appeal to the side of the human creature that craves certainty, likes to be elevated by mysteries, dislikes disagreeable truths, and clings to the flattering idea that we are not just remarkable animals, but chosen creatures." How fair is this characterization of the religious enterprise? If human beings are taught to accept mysteries, how does that square with the craving for certainty? Why the phrase "elevated by mysteries"? Mysteries may intrigue, frustrate, intimidate, but in what way do they elevate us? Don't they mark the limits of what we know? And what should we make of the disagreeable truths of a religion that says that we are born in original sin, that life is a vale of tears, that salvation is uncertain, and that we do not know whether we are saved or damned? In *Nature and Destiny of Man*, Reinhold Niebuhr characterizes "[m]odern man [as the possessor] of an essentially easy conscience; and nothing gives the diverse and discordant notes of modern culture so much harmony as the unanimous opposition of modern man to Christian conceptions of sinfulness" (23).

What religion does Haack have in mind that eschews disagreeable truths, or, for that matter, disagreeable falsehoods? In making her case, she looks for egregious examples of religious or theological thinking. "Theological responses to the Problem of Evil can be downright scary." She cites Richard Swinburne who finds himself "fortunate if the natural possibility of my suffering if you choose to hurt me is the vehicle that makes your choice really matter....[My] good fortune is that the suffering is not...pointless." To which she responds: "As it happened, the same day I read this I also read an article by a woman who had been raped, sodomized, terrorized, beaten and left for dead by a gang of thugs. Can you accept that she has benefited by having been the vehicle that made her attackers' choice really matter? I can't" (290). Leaving aside the mystifying syntax and the puzzling elisions in the citation, the reader may wonder why, among all the possible examples of theological thinking about the problem of evil, Haack chose the passage from Richard Swinburne. In fairness to Swinburne, it is difficult to infer from Haack's quotation of the passage what he means by the value of suffering. She might have cited Aeschylus who at the beginning of a long literary and moral tradition spoke of learning wisdom through suffering. If Haack is seriously interested in the religious response to the Problem of Evil, why didn't she address the thinking of Augustine and Thomas or the poets Donne and Herbert, or modern theologians such as Niebuhr and Tillich? In the *Evolution-Creation Struggle*, Michael Ruse, philosopher of science and self-declared Darwinist, responsive to the claims of religion, speaks of "the strident and intentionally hurtful atheism of Dawkins and his kind. Who would want to agree with such a person, even about science" (273)? He counsels "atheists like Dawkins and Coyne" (he might have included Haack) to "consider taking a serious look at contemporary Christian theology...rather than simply parroting the simplistic schoolboy travesties of religion on which their critiques are founded" (274). You don't have to be in agreement with a religious vision of life to feel the obligation to represent it accurately, especially if you are committed to the scientific ideal of objective truth.

There is much to object to in religious practice: self-righteousness, dogmatism, the idea of chosenness, intolerance of and cruelty (indeed, murderous cruelty) to infidels, but they do not encompass the totality of religious experience, nor are these vices confined to religion. These vices also apply to secular dogmas that eschew religion—for instance, Marxism or Nazism? Marxism calls itself a science. Dawkins and Haack would undoubtedly respond that it is a false science. Fair enough, but then we might want to ask ourselves whether the banishment of religion from a society might not deprive those suffering the tyranny of a false science of an alternative space of freedom in which their spiritual needs can be addressed.

When religion and science compete in their explanations of the origins of the universe, the origins of life and the material workings of the universe, religion loses. But in its symbolic aspect, religion provides satisfactions of needs for many people (educated and uneducated) that science can never provide. Religion is the source of the language of the spiritual and moral life that has no counterpart in the language of science. The sense of guilt, moral despair, the experience of suffering, contrition, conversion, redemption: no neuro-scientific explanation of the behavior of molecules in the brain can address and speak to the human beings who undergo these experiences. The physical resurrection of Jesus may have never taken place, but the resurrection need not be construed as a physical event. The higher Biblical criticism has taught us that the resurrection may be understood symbolically as representing changes in the inner life of a person as he or she is lifted from the depths of despair to an affirmation of life. When science dismisses a question as vain because it is in scientific terms unanswerable, it obtusely ignores the human need to ask these questions—such as why the innocent suffer, so memorably asked in *The Brothers Karamazov*. Imagine a world suddenly deprived of the language of religion, a world ruled exclusively by science and its explanations. Only a very small fraction of humanity would have access to a language that cannot enter into the emotional and spiritual lives of most people. In a review of a book by Stephen Rose in the *London Review of Books*, Ian Hacking endorses his suggestion that "there are two languages for describing connected events, a neural language and a language of experience. The second language is one that we share with many other people—it is not a private language at all" (20). Religious language is part of our vernacular, the language of human experience, so it does not matter what our religious convictions are, or even if we have none; it is ineradicably part of our lives. Religion too is a source (not the only source, to be sure) of community. As Durkheim has shown, when people congregate to worship, they are in a sense worshiping the community, an entity greater than any individual within it.

And what of non-observant secular persons like me? Don't we have spiritual and moral lives without religion? Yes and no. Certainly institutional religion, let alone a literalist belief in Biblical narrative, is not a spiritual requirement for everyone. (In many instances, it may even be a detriment.) But secular lives have been inspired and nourished by the arts, music, and literature: Shakespeare, Milton, Dante, Tolstoy, Dostoevsky, Bach, Handel, Michelangelo, Raphael, the list is very long, and so much of the literary, artistic, and musical imagination has been "parasitic" on the ideas and language of religion. When we despair and are lifted out of despair, when we experience the sentiment of love and charity, when we invoke a sense of righteousness and justice, however disaffiliated we may be from religious institutions and practices, the legacy of the religious imagination is abiding. Despite the frustrations of a militant atheist like Dawkins, who believes that the world would be better off without religion, it persists—and not because of the incorrigible childish naiveté or stupidity of masses of people, but because it represents a permanent adult need that the Enlightenment cannot trump. William James understood this without himself believing in the Absolute. Though James allies himself with the secular humanist view that "man is the measure," he resists the idea that human experience "is the highest form of experience extant in the universe." He "believe[s] rather that we stand in much the same relation to the whole of the universe as our canine and feline pets do to us" (199). The persistence of religious or spiritual aspiration is a fact that cannot be overcome by rationalist contempt. (Note: Rationalism is not always rational, nor does science exhaust all the possible uses of reason.) Which is not to say that the literalist credulity of religious fundamentalism should not be combated as false to the facts or that religious self-righteousness should not be resisted. Perhaps the most effective weapons against religious pathologies are homeopathic: the humility and rationality in the religious tradition itself.

As we have seen, religious sentiment persists even in Dawkins. Michael Ruse, in fact, speaks of evolution as having a religious as well as scientific aspect. E. O. Wilson, he notes, believes that human beings need religion to function, so he proposes Darwinism as a religion. "Religion writ large," he calls it. But what does that mean? What are features of Darwinism that would make it a religion in a humanly satisfying way? What are its institutions and rituals and what are its consolations? And how does it contribute to a sense of community? Does Darwinian religion assume a real knowledge of evolutionary biology, the kind of knowledge that biologists possess? In which case, can it possibly become the religion of non-scientists? Is it then destined to be the religion of the scientific few? Or are the scientists ordained to be priests of a new religion, privy to arcane knowledge to which the laity has no real access, but to which they must submit? And if so, for the many, Darwinism would be based simply on faith, not reason or evidence.

Dawkins's atheism is militant and intemperate. Daniel Dennett, no less an atheist, tries to come across as moderate, who, in the spirit of science, wishes to understand religion as a natural phenomenon. The very title of his book *Breaking the Spell: Religion as a Natural Phenomenon* (2006) is not promising, for it suggests a disposition to discover facts that will confirm his personal view of religion. Of course, he would deny such a judgment. By breaking the spell, he means two things: (1) breaking the religious taboo against scientific inquiry into the nature of religion and (2) breaking the spell of religion itself. He urges those who believe that religion is immune to scientific examination "to set aside our traditional reluctance to investigate religious phenomena scientifically, so that we can come to understand how and why religions inspire such devotion" (28). What can be more reasonable than to want to learn more about our deepest convictions? The "traditionally reluctant" person, however, might well respond by wondering about the good faith of Dennett's claim for the openness of his scientific inquiry when he declares his intention to break the spell of religion. Does he mean to demystify it in order to increase the ranks of atheists? One has the impression as one reads on of equivocation between the desire to understand and explain and the desire to expose religious faith as based on illusions. But that is not all. Dennett assumes that all resistance to his enterprise is a resistance to scientific inquiry. The "scientific" study of religion is hardly a new thing. The higher critics of the nineteenth century believed that in their study of Biblical literature they were engaged in scientific inquiry, as did twentieth-century anthropologists such as Edward Evans-Pritchard, Edmund Leach, and Mary Douglas. The suspicion of bad faith is aroused, for instance, when Dennett cites and comments on Evans-Pritchard's classic study of witchcraft.

> In his classic *Witchcraft, Oracles and Magic Among the Azande* (1937), the anthropologist Edward Evans-Pritchard vividly describes these proceedings, observing how the shaman cleverly enlists the crowd of knowing onlookers, turning them into shills, in effect to impress the uninitiated, for whom this ceremonial demonstration is a novel spectacle.
>
> [Here is the passage to which Dennett is referring] It may be supposed, indeed that attendance at them has an important formative influence on the growth of witchcraft beliefs in the minds of children, for children make a point of attending them and taking part in them as spectators and chorus. This is the first occasion on which they demonstrate their belief, and it is more dramatically and more publicly affirmed at these séances than in any other situation. (148)

The disconnect between quotation and commentary is striking. The point of view in the quoted passage is that of the spectators; the point of view in Dennett's comment is that of the manipulative shaman. Nowhere in the passage is there any suggestion of the shaman turning "knowing on-lookers," into "shills." The shaman is not even present in the quotation. Evans-Pritchard's tone and manner are objective and *scientific*, Dennett's tone and manner irresponsible and demagogic. He shows no interest in an objective study of witchcraft.

By science, Dennett means evolutionary theory, and given the discourse of genes, memes, and their replication that he employs in addressing religion, it is by no means unreasonable to be skeptical about the fruitfulness of the inquiry. Evolutionary theory addresses the question of why phenomena survive or disappear. "What pays for religion" is a refrain throughout the book. "Any such regular expenditure of time and energy has to be balanced by something of 'value' obtained, and the ultimate measure of evolutionary 'value' is *fitness*: the capacity to replicate more successfully than the competition does. (This does *not* mean that we ought to value replication above all! It means only that nothing can evolve and persist for long in this demanding world unless it somehow provokes its own replication better than the replication of its rivals.)" (69-70). The commercial metaphor (what pays for) is not fortuitous: "...this metaphor nicely captures the underlying balance of forces observed everywhere in nature, and *we know of no exceptions to the rule.*" Dennett knows that he is "risking offense." "Abhor the language if you must, but that gives you no good reason to ignore the question." But language is not incidental to knowledge and understanding. As is the case with the Barashes' insensitive use of language with respect to their approach to literature, Dennett's indifference to the quality and value of language in describing religion is not a matter of the breaking of a taboo, it is a betrayal of the subject under study: "people may well depend on religion, the way we all depend on the bacteria in our guts to digest food" (310). You don't have to be allergic to scientific inquiry to question the "payoff" (to adopt Dennett's idiom) in crude reductionist approaches to a subject. He is disdainful of the humanist complaint against reductionism as an expression of anti-scientific prejudice. But Dennett devotes so many pages to anticipating objections from those who might resist his inquiry that the reader may wonder about his own defensiveness. Is he saying that any objection to his approach is immune to reasonable criticism? For instance, one may not be objecting to the idea that religion can be the object of scientific study in wondering whether a particular approach to the subject qualifies as science.

Evolutionary theory is a science, but its applications, where there is little or no experimental evidence, may not be science at all. Much of the "science" advanced in the book gives answers to questions that have not been proved, but which Dennett believes have a good chance of being proved. After asking a series of questions such as why people are comforted by religion in their suffering, why religion allays our fear of death, why they are satisfied by religious explanations of things that can't otherwise be explained and why religion encourages group cooperation, Dennett promises to provide *the best current version* of the story science can tell of how religions have become what they are: "The main point of this book is to insist that we *don't* yet know—but we can discover—the answers to these important questions if we make a concerted effort" (103). Dennett attributes the explanatory function to what he calls "the intentional stance," the inclination to look for agency in events that seem to have come from nowhere. (Do I deserve his rebuke that I suffer from "premature curiosity satisfaction" if I am unimpressed by a "discovery" that is already common knowledge?) It is not science that arouses resistance, but the unconvincing portentousness with which Dennett claims that he is doing science.

Dennett's definition of religion (admittedly a very difficult word to define under any circumstances) is reasonable. "Tentatively, I propose to define religions as *social systems whose participants avow belief in a supernatural agent or agents whose approval is to be sought*" (9). What is left out, however, are beliefs in a spirit immanent in the natural world at odds with the materialist conception of nature necessary to scientific progress. It is a little odd that though he considers religion as a natural phenomenon, he does not consider religions that have as their object nature herself. Dennett's definition is a tentative one. As it turns out, his quarry is not only religion in the restricted sense of belief in supernatural agents, but rather fanaticism, whether its focus is ethnicity or political ideology. Thus in addressing the toxicity of "zealots of all faiths and ethnicities" (256) in the last century he cites the Serbian battle for Kosovo in which both religious and ethnic loyalties played a part as well as the Taliban's desecration of Buddhist monuments and the internecine warfare between Hindus and Muslims. He could have mentioned the even greater enormities of Nazism and Stalinism, which could be construed as the result of religious fanaticism only by a redefinition of religion to include secular faiths. (There was nothing supernatural about Hitler or

Stalin or Mao or Pol Pot.) The greatest atrocities of the last century were not committed by religious zealots, but by political ideologues for whom religion was anathema. In the instances of secular totalitarian society, religion could be a site of resistance to tyranny. Dennett's lumping together of phenomena that don't belong together has none of the precise distinction making one expects of the scientific intelligence.

So much of Dennett's book is speculation and moralizing, much of it having no need of evolutionary theory as a support. Who of rational persuasion would want to quarrel with his argument against unquestioning obedience to unjust authority or with his call to moderate adherents of Islam to resist the destructive extremism of its fanatical followers? But we don't require a scientific perspective to hold this view. Dennett's insistence on his scientific approach to religion, it seems to me, goes beyond normal scientific desire to understand the world. Like Dawkins and other colleagues in the field, he would like to replace religion with science, though he is sufficiently realistic to acknowledge the difficulty, if not impossibility, of its universal acceptance. What he fails to realize is that this explains much of the resistance to his version of scientific inquiry. Anthropological and higher critical study of religion does not provoke comparable resistance. Science does not replace the natural world that it studies. The "science" of Dennett and his colleagues in the instance of religion has the ambition to replace the object of its study. The neo-Darwinists discredit their atheism (a legitimate creed) when they try to dress it up as science.

* * *

Darwinists are not alone in the community of scientists in wanting to find spiritual fulfillment in their pursuit of scientific truth. The string theorist Brian Greene is an interesting case in point. In *The Fabric of the Cosmos: Space, Time and the Texture of Reality*, he recalls his teenage "encounter" with Camus's *The Myth of Sisyphus*. He was startled by its beginning: "There is but one truly philosophical problem and that is suicide… whether or not the world has three dimensions or the mind nine or twelve categories comes afterward" (3). The question of suicide is, of course, the question of whether life is worth living. The question for Camus grows out of the despair one may experience in daily life. As a grownup scientist, Greene came to understand the appeal of this view: "it's easy to be seduced by the face nature [i.e., ordinary life] reveals directly to our senses." But Greene the scientist resists the seduction, because "the overarching lesson that has emerged from scientific inquiry over the last century is that human experience is often a misleading guide to the true nature of reality. Lying just beneath the surface of the everyday *is* a world we'd hardly recognize" (5). Gaining access to that world, Greene asserts, would lead to an affirmative answer to Camus's question. And so, in an affirmative spirit, Greene pursues his inquiry into the unseen fabric of the cosmos. The discovery of scientific laws hidden by "the face of nature" is the *via media* to a kind of transcendence of ordinary phenomenal life that is comparable to religious experience.

The unaddressed problems packed into Greene's affirmative answer make it difficult not to view his answer as glib. In what respect is everyday experience "often a misleading guide to the true nature of reality"? Is a person who is struck down by a fatal or severe chronic disease being misled by his experience? Is the effect of a tragic love affair or incurable loneliness an illusion to be remedied by knowledge of the ten or eleven dimensions that string theory has discovered? Everyday experience may contain happy events such as marriage or the birth of a child or success in one's career, and Camus's question on the basis of that experience might yield an affirmative answer, depending of course on the temperament of the person. By distinguishing between the illusory face of nature and its hidden physical laws, Greene *appears* to be engaged in a religious enterprise, but what the unseen of physical science lacks is precisely what religion possesses, an insurmountable sense of mystery. The laws of physics may be invisible to the senses, but when discovered they are known to reason and the sense of mystery vanishes. The laws are made visible in mathematics to those trained to read and understand mathematics. The search for experimental validation and mathematical confirmation, not faith, is the force that drives Greene and other physicists. He may experience a sense of wonder at the discoveries he has made, but what he seeks to overcome is mystery. He wants to replace the unknown with the known. What the physical laws of nature fail to illuminate are the arbitrary and chance events that make for unhappiness, happiness, suffering, and joy. How will knowledge of the laws of nature, should we acquire it, console us when we suffer the vicissitudes of everyday life? Religion may not solve the problems of the human condition, but it speaks to "the face of nature" in which the laws of physical nature play no part.

Greene's personal answer to the question of whether life is worthwhile is affirmative, having found the value of his existence in his vocation as a scientist. But what does science offer in the way of spiritual satisfaction to non-scientists, that is, to those of us who have no direct experience of the making of scientific discoveries and only minimal understanding of them? We may trust (have faith?!) in what scientists tell us about the physical universe, but we will not have experienced the excitement and happiness of having made the discoveries. It is not the hidden laws of nature that are the source of Greene's conviction that life is worthwhile, but rather his sense of vocation. His scientific work is part of his everyday reality. Others will find a similar conviction in other vocations. That sense of vocation, whatever that vocation may be and not the hidden laws, is part of our daily experience and gives meaning to our lives. Greene may have found happiness in his vocation as a scientist; others may find contentment or its opposite in other vocations. Camus's question belongs to an order of *real,* not illusory, everyday experience to which the hidden laws of nature are irrelevant.

The sense of vocation (translated as a calling) suggests a religious inspiration, and we can appreciate the feeling of life's worth that Greene and his colleagues find in their work, but is he entitled to a religious idiom when he speaks of the seductions of everyday life and the reality that is hidden behind it? He cites the example of Einstein, who told the philosopher Rudolph Carnap that "the experience of the now means something special for man, something different from past and future, but this important difference does not and cannot occur within physics." And in a condolence letter to the widow of a friend, Einstein wrote that "in quitting this strange world he has once again preceded me by just a little. That doesn't mean anything. For we convinced physicists the distinction between past, present and future is only an illusion, however persistent" (*New York Times* Jan.1, 2004, A 23). Einstein was not providing the widow with the consolation of an eternal afterlife. The eternal truth of physical science is an abstraction best grasped through the abstractions of mathematics. So the question posed by Greene is the following: Should we correct our sense of time according to scientific knowledge, or should we allow our intuitive, phenomenological experience (an illusion according to physics) to prevail?

Apart from the difficulty of anyone but a scientist who may know what it is to live in this time-cancelled way, there is the question of what would be gained and what lost if we gave up our sense of past, present, and future. The scientific erasure of phenomenological or experiential time would take the drama out of life. It would devalue the senses, memory and the emotional life. History and art would disappear. Greene speaks of his effort to "bring my experience closer to my understanding. In my mind's eye, I often conjure a kaleidoscopic image of time in which, with every step, I fracture Newton's pristine and uniform conception.

And in moments of loss I've taken comfort from the knowledge that all events exist eternally in the expanse of space and time, with the partition into past, present and future being a useful but subjective organization." "Eternally in the expanse of space and time" is the language of religion, but without its content. It is not at all clear what spiritual comfort can be derived from such knowledge. Moreover, Greene knows that even the scientist cannot exist in scientific reality: "Regardless of our scientific insights, we will still mourn the evanescence of life and are able to thrill to the arrival of each newly delivered moment." He goes on to say: "The choice, however, of whether to be fully seduced by the face nature reveals directly to our senses, or to also recognize the reality that exists beyond perception, is ours." Of course, the fact of nature is seductive and illusory only if we accept the view that only the physicist's conception of time is real. Are *mourning* and *thrilling* illusions? In any event, I doubt whether the choice that Greene proposes is a real choice for anyone except the physicist. What is interesting about Greene's aspiration is his need to find a religious fulfillment in the work of science. I am skeptical about whether eternity as conceived by physics qualifies as religious or spiritual experience. I am convinced, however, that the eternity, which Greene strives for is available, if available, only to the community of physical scientists.

Tolstoy saw in the progress of science not the promise of greater happiness for humanity, but rather an occasion for despair: "[T]he man immersed in modern culture, in the unending growth of knowledge, can become 'tired of life,' never 'sated,' for he cannot appropriate more than a tiny fraction of what has been made available to him in modern civilization. In the face of this infinite wealth, the death of the individual has become an abrupt and pointless interruption of his potential development. 'Progress' has made death a contradiction, and thus the boundless increase of knowledge itself has lost its meaning" (Ringer, 234). Should this prospect make us despair? Only if we believe that the possession of total knowledge (the realization of the dream of a final theory) would make for happiness? In responding to Tolstoy's despair, Max Weber evokes Plato's cave, where one of the prisoners accustomed only to the shadows cast upon the wall, freed from his chains, turns to the opening of the cave to face the sun of truth and beauty. The prisoner becomes the philosopher who in contemplating the good and the beautiful achieves a happiness available only to the philosopher. The truth that Plato's philosopher promised is moral, what the eighteenth-century writers would call aesthetic knowledge and what a twenty-first-century physical scientist would call a final theory of the universe. But the question remains whether the happiness (the sense of life's worth) depends on the possession of knowledge, as Greene apparently believes. The capacity for happiness may be a matter of instinct and temperament. Tolstoy knew this in his role as novelist. Think of Gerasim, the peasant servant in *The Death of Ivan Ilych*, whose wisdom and contentment have nothing to do with the possession of boundless knowledge. Nor are his wisdom and contentment disturbed by the prospect of mortality. Gerasim may be Tolstoy's aristocratic fantasy about peasant life, but there are people in real life who have Gerasim-like capacities. In any event, for those of us who need the feeling that our lives have meaning and worth, the route to follow (again for most of us) is not modern science. It is rather the moral or religious or artistic imagination, which may or may not exploit the discoveries of science. I don't mean to suggest that science as a vocation is not meaningful, but rather that meaning does not necessarily spring from the discoveries of science. The meaning of individual lives is not to be found in reproduction (*pace* Richard Dawkins) or in the eleven dimensions of string theory (*pace* Brian Greene). Science, an extraordinary and powerful achievement of mankind, is not salvation.

The distinguished biologist Peter Medawar makes a useful distinction between scientific messianism, which derives from the radical Enlightenment belief in the prospect of a rational utopia, and scientific meliorism. The messianic dream is that a free "peaceable, loving, cooperative world" will come into being as a result of the material progress produced by science. In *Advice to a Young Scientist,* Medawar confesses to a certain affection for this view, which he believes proceeds from "a lovable human trait," but he nevertheless finds "precious little incentive to believe that the 'doctrine of original virtue,' the contrary of the doctrine of original sin is true" (103). The scientific messianists, drawing inspiration from a belief in the natural goodness of human beings, see the solution of all of life's problems in the dispelling of ignorance about the sources of evil. The intellectual and scientific mastery of the world will produce a utopia at once rational and moral. I'm not so sure that this view reflects "a lovable human trait," since messianism, whether scientific or religious, is often accompanied by tyrannical arrogance about where the truth lies. The proverb about the road to hell haunts the messianic dream. In any event, Medawar in the spirit of Karl Popper's open society declares himself on the side of scientific meliorism and the never-ending task of improving an always imperfect world. He has the scientist's hopefulness in contrast to what Stephen Graubard has called "the habitual despondency of literary humanists," though there is enough of the despondent humanist in him to side with Voltaire's *Candide* against Leibnitz's optimistic theodicy. "All is not well. This is not the best of all possible worlds." Despondency, however, is not the right word for Medawar, for knowledge of the miseries of life does not prevent scientists as well as the rest of us from trying to make the world better.

Max Weber made perhaps the most persuasive case for the necessity of specialization in the sciences, but he also understood its costs. He understood that, in the words of Fritz Ringer in his intellectual biography of Weber, "science could give no definitive answer to questions about 'how we should live." And he "held a deliberately modest view of science, arguing that it required an independently grounded presupposition about what was worth knowing." In the concluding chapter, Ringer expresses his appreciation for his view of the relations between science and religion in a manner that speaks to my own sense of the fraught relationship between science and the humanities.

> Above all, I am enthralled and converted by Weber's comprehensive history of human *reason* in the evolution of human culture. The gradual "removal of magic from the world," which began in religious thought, came to assert itself over all kinds of obstacles, until it encompassed the systematic rationalization of *life conduct* as well as of thought. Weber experienced the ultimate outcome of that process as a tragedy, because he was not as "unmusical" with respect to religion as he believed. And he allows us to be moved by the sharpening conflict between science and religion that was the trauma of his time. He deeply felt the tension between the unconditional imperative of Tolstoy's "cosmic love," even while remaining committed to science—and to the view of religion as the "sacrifice of the intellect." (254)

What I find admirable in this passage is the poise with which Weber contemplated the cultural situation of his time, reflected in the poise of Ringer's description. We need not follow Weber in all his analyses to embrace the kind of balance that he achieved as a model for our own time.

* * *

Every discipline has vices toward which it tends. Religion may, indeed too often does, breed fanaticism and cruelty to the infidel, science, a cold neutrality toward destructive technological consequences of its discoveries, aestheticism, an indifference to the ethical life, ethics, a censorious or repressive view of vital (amoral) energies. But none of these vices defines the discipline. What a constructively critical view of a discipline requires is vigilance about the extent to which a vice may take over the discipline at any time. What needs to be cultivated is a critical sense of both the powers and limitations of a discipline.

If we respect the separate integrity of the disciplines as I believe we should, we might be able to think our way to a negotiated resolution of the conflict between the evolutionists and their opponents, the creationists and the advocates of intelligent design. Here is a peace proposal. First, what needs to be accepted by both sides is that victory is usually temporary and illusory, for it produces in the defeated a resentment that can always erupt in the future. In a negotiated settlement each side makes concessions to the other and might gain advantages as a result. Religious fundamentalists must concede the terrain of science to evolutionary biology (or Darwinism), which is a theory supported by a vast body of data. It should be noted that the theory exists in different versions that in certain respects are at odds with each other. The controversies within evolutionary theory should be addressed. As a science evolutionary biology deals in probabilities, not certainties, which does not diminish its scientific character. Indeed, the conviction of certainty is inimical to the scientific enterprise. In return for this concession, religion would be allowed to enter the curriculum of the public schools as a legitimate subject for study. A course in religion would be ecumenical in its hospitality to Christianity, Judaism, and Islam. Its perspective would be historical, but it would also allow for a critical consideration of the relevance of the moral and spiritual claims of religion to contemporary life. The Bible, for instance, would be read with the same seriousness with which students read or ought to read Shakespeare and Milton. A course on religion would be sensitive to the sensibilities of the devout and skeptical. The devout may find their faith reinforced and the skeptics their doubt intensified, or conversely the devout might find themselves questioning their faith, while skeptics might find inspiration in what they previously doubted. What would be encouraged is a critical openness to religious experience. The course or courses would proscribe proselytizing and of necessity be elective.

I can anticipate objections from both sides. The advocates of the separation of church and state want to keep religion entirely out of the public space (the schools belong to that space) out of an understandable anxiety that it would become an opportunity for proselytizing. But our Constitution guarantees religious liberty, which allows for its expression in the public as well as the private arena. What the constitution does not permit is religious coercion in the form of proselytizing. We have the experience of courses in religion at the college and university level to go by in setting the standards for non-proselytizing courses in the public schools. Religion in its various expressions has been part of human culture. Why then should it not be part of the formal education of young people? Of course, the recruitment of the right faculty would be essential to the success of the subject. I can think of no better source of recruitment than modern divinity schools, given their openness to the variety of religious experience. Whether an instructor is Protestant, Catholic, Jew, Muslim, agnostic, or atheist, he or she would have to be respectful of the subject. The fundamentalists, on the other side, may find such a course too weak a brew to quench their thirst, for it would not guarantee confirmation of their views or the conversion of non-believers. But they will be getting something they never had before, the introduction and validation of religion as a subject in the public curriculum. Some settlements are compromises that reduce the advantages of both sides in the conflict. The effect of this proposal, I believe, would be the enhancement of the interests of both sides. What needs to be disentangled in the negotiation in order to achieve the desirable result is the confusion created by the attempt to conflate science and religion.

3

Reinventing Ethics

What happens when Darwinism becomes the determining force in moral philosophy? We have the remarkable example in the work of Peter Singer. He has been variously characterized hyperbolically as "the most controversial," "the most influential," and "the most notorious philosopher" in the world. His books cover a wide range ethical issues, from animal liberation to euthanasia (including infanticide) and from globalization to the workings of George W. Bush's mind. The controversy, influence, and notoriety come from positions he has taken on the most sensitive issues of life and death with a candor that one rarely finds in the work of a respectable philosopher. Whether or not one is in agreement with his views (I am not, as will soon become evident), they are argued with a provocative clarity and rigor that must be taken seriously. A radical advocate of animal rights, he wants to dissolve the species distinction between human beings and non-human animals. An advocate of euthanasia and infanticide when necessary, he substitutes for the commandment that honors the sanctity of *human* life one that sanctifies the quality of life. In both instances, human beings are denied a privileged place in the world simply by virtue of being human.

If, as Copernicus demonstrated, the universe is not earth-centered, and if, as Darwin has shown, human beings are not the separate creation of God, there can be no justification for providing human beings with a separate ethical system, often at the expense of non-human animals. Singer calls this speciesism and analogizes it to racism. In *Rethinking Life and Death: The Collapse of Our Traditional Ethics* (1994), he argues that in their susceptibility to pain and pleasure, in their capacity for consciousness and even in certain instances a rudimentary self-consciousness, animals deserve to be treated with the same consideration as human beings. He does not deny the superior capacities of human beings, but he views the differences between human and non-human animals as differences in degree and not kind. He provides the following memorable account of the gorilla Koko:

> She communicates in sign language, using a vocabulary of over 1,000 words. She also understands spoken English and often carries on "bilingual" conversation, responding in signs to questions asked in English. She is learning the letters of the alphabet, and can read some printed words, including her own name. She has achieved scores between 85 and 95 on the Stanford-Binet Intelligence Test.
>
> She demonstrates a clear self-awareness by engaging in self-directed behaviours in front of a mirror, such as making faces or examining her teeth, and by her appropriate use of self-descriptive language. She lies to avoid the consequences of her own misbehaviour, and anticipates others' responses to her actions. She engages in imaginary play, both alone and with others. She has produced paintings and drawings which are representational. She remembers and can talk about past events in her life. She understands and has used appropriately time-related words like "before", "after", "later", and "yesterday".
>
> She laughs at her own jokes and those of others. She cries when hurt or left alone, screams when frightened or angered. She talks about her feelings, using words like "happy", "sad", "afraid", "enjoy", "eager", "frustrate", "made" and, quite frequently, "love". She grieves for those she has lost—a favorite cat who has died, a friend who has gone away. She can talk about what happens when one dies, but she becomes fidgety and uncomfortable when asked to discuss her own death or the death of her companions. She displays a wonderful gentleness with kittens and other small animals. She has even expressed empathy for others seen only in pictures. (181)

Why then should animals (Koko is unusual but not alone in these capacities) not be accorded the same ethical consideration as human beings? My answer, not Singer's, is that the differences between human and non-human animals in intelligence, linguistic and symbolic capacity, consciousness and self-consciousness may be a difference in degree, but of such an order of magnitude that quantity becomes quality. To say that non-human animals have 98 percent of DNA in common with human beings tells us nothing of the effective intellectual superiority of human beings. No matter, Singer might respond, for he has a deeper reason for his anti-speciesism. "Whether or not dogs and pigs [animals without Koko's capacities] are persons, they can certainly feel pain and suffer in a variety of ways, and our concern for their suffering should not depend on how rational and self-aware they might be" (182). What right do we have, for instance, to extract the heart of a healthy and vigorous baboon for the purpose of transplanting it to a sick and aging human being? The reason for our willingness to do so lies in what Singer characterizes as the Old Commandment: "Treat all human life as always more precious than any nonhuman life." Singer proposes the following substitution: "Do not discriminate on the basis of species." What follows is the refusal to value a healthy non-human animal above a sick human animal. (We need always to be reminded of our animal essence.) Singer's view is based in Darwin, who wrote: "Man in his arrogance thinks himself a great work, worthy of the interposition of a deity. More humble, and, I believe, true to consider him created from animals" (169-70). The quotation and the conclusions that he draws from it should make us realize their distance from the humanism of the Renaissance and the Enlightenment, which in its taxonomy of species remains derivative of Christianity and the belief in a hierarchy of separate kingdoms of creation. Singer means to draw out the anti-*human*ist implications of Darwinism and in the process means to show the inhumane aspect of humanism in its indifference to the suffering of non-human animal life.

When Singer speaks of the intelligence of animals, of their susceptibility to suffering, and of their capacity for empathy, his examples are apes and chimpanzees, animals that most closely resemble human beings. We sympathize with them because they are like us. Which means that human beings still provide the standard for what is most valuable in life. But Singer's embrace of animal life is ecumenical and therefore must be extended to creatures that do not possess Koko's capacities. How would Singer have us decide when confronted with a choice between an intellectually disabled human and a healthy rat? Would it ever be ethical, according to his lights, to perform experiments on rats or mice in order to find cures for human diseases? I do not know how Singer would answer the question, but I do know how his doctrine would require him to answer it.

Here are practical questions for Singer. How intellectually disabled would a human being have to be to lose out in a life and death situation to a healthy baboon? What if the IQ of an intellectually disabled human being were marginally higher than that of a healthy baboon or chimpanzee? How would that affect the decision to extract its heart? Does a high baboon score among baboons belong in a different category from a low human score, in which case we are speaking of a species difference? Singer's case against speciesism does not rest exclusively on the rudimentary rationality of the higher animals. He is legitimately concerned about the suffering human beings inflict upon animals, for instance, in the slaughtering of cattle. Isn't it then possible for human beings to make every effort to prevent cruelty to animals and to alleviate gratuitous suffering and still believe that humans and sub human animals belong to different species and should be treated differently? In the instance of transplanting the heart of a baboon to a human being, the baboon would be put to sleep before surgery and not suffer.

What should we make of the analogy of speciesism to racism? Racism has many manifestations, one of which is the taboo against interbreeding and marriage between members of different races. It is perhaps racism's deepest taboo. If the analogy holds, speciesism would entail a taboo against interbreeding between, say humans and chimpanzees, which anti-speciesists would want to combat, just as an anti-racist would fight against proscribing marriage between a white person and a black person. Singer takes up the challenge or appears to take it up. He cites Richard Dawkins's discussion of ring-species, in which he points out that though the herring gull and the lesser black gull seem to be "clearly distinct species," they are on a continuum with intermediate species that have died out. "So," Singer concludes, "if we cannot interbreed with chimpanzees, this is merely due to the deaths of the intermediate types. In any case, why do we assume that a human being and a chimpanzee could not produce a child? It is true that there is a difference in the number of chromosomes, chimpanzees having 48 and humans 46. But siamangs and lar gibbons—two distinct species [sic!] of ape living in Malaysia and Indonesia—have interbred, despite the fact that they have different numbers of chromosomes (50 and 44 respectively). So the possibility of human and chimpanzee interbreeding cannot be ruled out" (179-80). I'm sure I am not alone in finding this passage simply weird. But I will suspend incredulity and ask if there are any experiments to determine whether such interbreeding is possible. Would anyone want to perform such an experiment? Would it not be grotesque to do so? And let us say that interbreeding was possible, what would we do with the offspring? Is it reasonable to surmise that the half-human, half-chimpanzee might be a candidate for infanticide following Singer's own ethical code? The analogy between speciesism and racism is an absurdity.

The liberal Protestant theologian Reinhold Niebuhr provides an alternative and illuminating view of man's paradoxical status in the evolutionary process:

> If man insists that he is a child of nature and that he ought not to pretend to be more than the animal, which he obviously is, he tacitly admits that he is, at any rate, a curious kind of animal who has both the inclination and the capacity to make such pretensions. If on the other hand he insists upon his unique and distinctive place in nature and points to his rational faculties as proof of his special eminence, there is usually an anxious note in his avowals of uniqueness which betrays his unconscious sense of kinship with the brutes. This note of anxiety gives poignant significance to the heat and animus in which the Darwinian controversy was conducted and the Darwinian thesis was resisted by traditionalists....The obvious fact is that man is a child of nature, subject to its vicissitudes, compelled by its necessities, driven by its impulses, and confined within the brevity of the years which nature permits its varied

organic form, allowing them some, but not too much, latitude. The other less obvious fact is that man is a spirit who stands outside of nature, life, himself, his reason and the world. This latter fact is appreciated in one or the other of its aspects by various philosophies. But it is not appreciated in its total import. That man stands outside of nature in some sense is admitted even by naturalists who are intent upon keeping him as close to nature as possible. (1, 3-4)

Niebuhr focuses on man's capacity for self-transcendence without denying his roots in the animal world. Unlike religious fundamentalists whose view of man's place in the world derives from dogma, Niebuhr approaches the matter empirically by appealing to our experience of our place in the world. Without discounting the fact of man in nature, he resists reducing him to his natural or material aspect.

Darwinism has historically been an unreliable guide in ethical matters. Consider its pernicious consequences in Social Darwinism, in which the Struggle for Existence and the Survival of the Fittest become models for economic life. Darwin and his advocate Thomas Huxley cautioned against conflating evolution and ethics. In *Ethics and Evolution*, Huxley asserts that "no one is more strongly convinced than I am of the gulf between civilized man and the brutes.... Our reverence for the nobility of mankind will not be lessened by the knowledge that man is, in substance and in structure, one with the brutes" (quoted in Singer 172). Singer will have none of this. Huxley's motive was simply to "reduce opposition to the acceptance of Darwin's theory." We may want to discount the Victorian piety of the phrase "reverence for the nobility of man," but Huxley, I believe, was speaking from genuine conviction in affirming our qualitatively superior capacities to those of non-human animals, capacities we may abuse or misuse. The statement is consistent with his anti-Spencerian view (what came to be called Social Darwinism) that human life is not or should not be defined by the Survival of the Fittest, a doctrine that became a model for economic life. As a self-declared left Darwinian (as opposed to right [?] Darwinism of Wilson et al.), Singer dissociates himself from Social Darwinism. He is, nevertheless, undaunted in his attempt to extrapolate a benign morality from evolutionary theory.

Has he succeeded? Only if he is allowed to pick and choose arbitrarily from its full panoply. Once invested in a theory, you can't simply ignore what works against your personal desires. If Darwinism is the basis for rethinking ethics, then we will need to incorporate Natural Selection and the Survival of the Fittest in conceiving or reconceiving an ethical system. The very fact that human beings are essentially animals (and not angels) means that they too struggle for existence. They kill animals for food, resist or trim predatory herds to protect their crops and their domestic animals, experiment in the laboratory with animals so that cures can be found for human diseases. The discrimination on the basis of species can be viewed as natural according to the Darwinian dispensation. The ethical doctrine of non-discrimination of species must then have its source elsewhere—in the very (non-Darwinian) humanism Singer rejects. No non-human animal could possibly arrive at such a universal empathy for suffering creatures. Need it be said, Singer's very performance as a philosopher illustrates his *qualitative* superiority to Koko.

All living things have their origins in a single tree. We are not the result of separate creations, but, as Darwin himself has taught us, evolution is also a drama of variation and diversity. Separations among species arise from that single tree. We do not have to subscribe to Christian fundamentalism to hold the view. What our humanism requires is not the radical revision of taxonomy that Singer advocates, but a rigorous avoidance of gratuitous discrimination against living creatures. We need to protect our environment and the animals that live in it. But as animals, we cannot abandon the interests that make for our survival and prosperity. And those interests require us to regard ourselves as a separate species while humanely considering the interests not only of other species, but also of our planet.

However vulnerable Singer's anti-speciesism may be, it puts him on the side of the angels in its compassion for animal life. It is the Dr. Jekyll side of his ethical system. Mr. Hyde appears in his argument for making discriminations on the basis of the quality of life. It starts with euthanasia for the very seriously ill sufferers in old age, who cannot bear the suffering (a view that one can claim for the ethics of compassion) and moves toward an advocacy of infanticide for infants up to twenty-eight days old, who, in Singer's view, have not yet achieved personhood. He sanctions euthanasia for the incurably sick, the deformed, and the disabled infant. He has gone beyond the pro-choice abortion argument that asserts the right of women to have control over their own bodies to a radical rejection of the doctrine of the sanctity of human life. What matters is not life itself, even human life, but its quality. It is therefore not surprising that Singer has acquired the reputation as the most dangerous philosopher alive, inviting the absurd comparison to the notorious Joseph Mengele of the Nazi concentration camp. I mention the comparison only to set it aside as an obstacle to a genuine and fair consideration of what Singer is up to.

There is something of a paradox at the core of Singer's philosophy. He is an egalitarian in considering the claims of human and non-human animals. Who is to say that our superior intelligence and complex consciousness, for instance, of the passage of time, makes us more deserving of survival than gorillas and pigs? But within the human realm (the word "species" is taboo when preceded by "human") discriminations are permitted between, let us say, the mature healthy adult and the sick just-born infant. The elevation of animals to a place where they share personhood with human beings goes along with a devaluation of human beings to a place where the sanctity of human life is trumped by invidious distinctions between human beings in matters of life and death based on the quality of life. Singer believes that he is following Copernicus and Darwin in their implicit rebuke to human pride. Man, the putative image of God, no longer is the center of the universe. He exists on a planet in a solar system that is a tiny part of the universe. He is no longer to be distinguished from the animal kingdom. We may not be the center of the universe, but we do not have to be the center in the Christian sense (man made in the image of God) to know that we have powers for good and ill that transcend those of other species. We alone of the creatures of the earth have the capacity to save or destroy the planet. Scientific discoveries made from the Renaissance to the present may have demystified the idea that we are made in God's image, but they have not diminished humanity's self-esteem. On the contrary, with the extraordinary advances in science, humanity's belief in its powers has only increased. Irony of ironies, in the wake of the decline of religion, a new humanism arises which affirms man, not God as the measure of all things. Writing as humanist and pragmatist, William James declares that "to an unascertainable extent our truths are man-made products" (159). To be sure, man-made truths don't have the solidity or certainty of divine truths and are particularly vulnerable to skepticism even within the circle of humanists and pragmatists. Nevertheless, the idea that science has ministered a devastating blow to the centrality of human beings in the world and even the universe is something of an illusion. Scientific progress only contributes to the sense of humanity's god-like powers. Humility, which a God-centered universe requires of human beings, has become a questionable virtue in the modern world. Indeed, there is something prideful in Singer's willingness to set himself and his fellow human beings as life and death judges of the quality of human life.

Singer speaks of the need for developing a coherent ethical system to replace the one that he believes is collapsing: an ethics based on the sanctity of human life. Is the Jekyll and Hyde-like appearance of his philosophy an illusion that conceals a deep consistency between anti-speciesism and infanticide? What they do have in common is a compassion for suffering creatures. Unlike Mengele, Singer is not advocating infanticide as an exercise in cold-blooded experimentation. The purpose of infanticide is to prevent a deformed or disabled infant from living a life of irremediable suffering. There is nothing in Singer's thought and career that suggests that his personal motives are anything but humane. But an ethical doctrine cannot simply be justified by its motives. We need also take into account its standards and its consequences, particularly its consequences. Consequences are at the heart of utilitarianism, and Singer himself puts himself squarely in the camp of the utilitarians.

In substituting the quality of life for the sanctity of human life, Singer has in effect introduced the Survival of the Fittest as an ethical category. What was a descriptive phrase in Darwin became an ethical imperative in Spencer. Singer would doubtless reject the charge that he has in effect reintroduced Social Darwinism. As a man of the left, he would not endorse Spencer's application of the Survival of the Fittest to the realm of economics, but though he makes a different application of Darwinism to the ethical realm, the effect is similar. Science can inform moral judgments by providing information about the consequences of our behavior, about its costs and benefits, but our moral actions as individuals and communities have their source in values derived from religion or from its legacy in the secular realm.

Though, as I have already said, there is no reason to believe that Singer's motives are anything but humane, the criterion or criteria by which judgments will be made under the dispensation he proposes are not humane. As humane creatures, we value health over illness, intelligence over stupidity, but we do not or should value healthy human beings over the sick or the intelligent human beings over stupid ones where life and death are at stake. Singer invokes Darwin, but Nietzsche haunts the argument. And indeed what Darwinism or certain versions of it have in common with Nietzscheanism, from different perspectives to be sure, is a kind of ruthless anti-humanism. Consider the following consequence of Singer's doctrine of infanticide. An infant born with the physical disabilities of Stephen Hawking can be terminated during the first twenty-eight days of his life. One can hardly know what the infant may become as he grows up. The chances, of course, are that he will not turn into a great physicist. But the possibility alone must arouse our resistance to turning infanticide into a doctrine. Indeed, our resistance should not depend on whether greatness lies ahead.

Singer has aligned himself with the sociobiologists and the evolutionary psychologists who, among other things, seek to ground ethical life in Darwinian science, though he views himself on the Darwinian left in contradistinction to the right-wing (?) bias of most sociobiologists and evolutionary psychologists. Ethicists should be willing to learn from biology what it has to teach us about our human nature. But first, we need to be confident that what it says about our nature is true knowledge and does not mislead us. Then we need to recognize that the ethical choices we make can never be derived from our uncertain knowledge of human nature. They belong to our freedom to enact or to resist what we believe to be our natural impulses. The reduction of *ought* to *is* is a philosophical sin. What is objectionable in Singer's philosophy are not his thoughts on the subjects he chooses to reflect upon, but rather his eagerness to turn those thoughts into doctrine. In fairness to Singer, this eagerness is checked by his scruples, his awareness of the danger that lies in what he is proposing. "The strongest argument for treating infants as having a right to life from the moment of birth is simply that no other line has the visibility and self-evidence required to mark the beginning of a socially recognized right to life. This is the powerful consideration; maybe in the end it is even enough to tilt the balance against a change in the law in this area. On that I remain unsure" (217). If Singer is unsure in so grave a matter, should he have indulged himself in making a sustained argument for infanticide, denying an infant a right to life only to leave the proposal in state of uncertainty in a few concluding sentences? I am reminded of the journalistic habit of placing corrections in inconspicuous places in newspapers after harm has been done in long inaccurate articles.

J. M. Coetzee's recent novel *Elizabeth Costello* brings Peter Singer to mind, though he is never mentioned in the book. The eponymous hero is a novelist with a passion for animal rights. In a series of lectures and exchanges with her audiences, however, she, unlike Singer, rejects all justifications for the killing, or for that matter the preserving, of animals on rational grounds. When she is challenged to specify the reasons for her championing of animal rights, she demurs: "I was hoping not to have to enunciate principles.... If principles are what you want to take away from this talk, I would have to respond, open your heart and listen to what your heart says" (82). Contrast this with what one of her interlocutors (a philosopher?) argues: human beings in possession of souls are in revolt against death, while "to animals, death is continuous with life" (109). Ergo, it is licit to kill animals. Not that Elizabeth Costello does not argue and even give reasons for her views, but she finds the whole rationalist philosophical tradition inimical to what she feels and believes. The possession of consciousness and rationality simply does not give one license to kill those who do not possess them. If Costello and Singer are both advocates of animal rights, they are opposed in the ways of their advocacy. Costello is a romantic, Singer a rationalist, who not only gives reasons but also finds in the creatures he values the possibility and the power of rationality. In Costello's radical version of animal rights, we have an implicit critique of Singer's rationalist argument for infanticide. Costello's romanticism is vulnerable. She is under attack from other characters who are given plausible arguments for their opposing views. But unlike, Singer, she cannot be accused of scientism. Advances in medical knowledge and technology are not the driving force in her ethical views. She does not have the ambition to inaugurate a Darwinian revolution in the ethical sphere.

It may be better to have an ethical code with exceptions and anomalies that make for a certain incoherence than to create an ethical system that is procrustean. The exceptions and anomalies allow for awareness that different situations require responses which may not fit together in a consistent scheme. There are of course circumstances in which euthanasia is a humane decision, even the only humane decision. I should think that a wish not to be sustained on life supports in a vegetative state would trump the doctrine of the sanctity of human life. But I would want the decision to remove those supports to be made with the clear evidence that the wish of the invalid had been expressed. Moreover, I would not want the sanctity of human life to be set aside from revulsion, experienced by the healthy person who has the power of decision, from the condition of the vegetative state. From a utilitarian perspective (that of Singer) I would be concerned with the precedent a life-terminating decision might set. We should mistrust all systems of ethics that do not allow for flexibility and exceptions. Kant, for instance, enjoins us always to tell the truth, but we may want to qualify the injunction by concealing the truth or even lying when truth telling does harm. We may even find ourselves in the predicament of not knowing whether the truth does good or harm and hesitate before acting precipitously. What we do not need is a Darwinian revolution in the realm of ethics that would coerce us into making decisions that would violate our humanity.

The Polish Nobel Prize-winning poet Wisława Szymborska illuminates what radically distinguishes humans from other animals in her poem "In Praise of Feeling Bad About Yourself."

> A jackal doesn't understand remorse
> Lions and lice don't waver in their course
> Why should they, when they know they're right
>
> On this third planet of the sun
> Among the signs of bestiality
> A clear conscience is Number One.

Of course, a clear conscience is no conscience at all. And there are beings that purport to be human who behave as if they have no conscience: they kill and massacre. They are no different from lions and lice, or more accurately, worse than lions and lice because their presumed humanity gives them a capacity for evil far exceeding that of subhuman animals. Lions and lice are beasts, but they are not under the sign of bestiality, another way of distinguishing the human.

Postscript: Singer may be an eccentric example of Darwinian ethical thinking, though I believe that in his idiosyncratic way he represents a tendency among neo-Darwinists to want to revise or at least substantiate our ethical categories in the evolutionary process. Daniel Dennett, a passionate philosophical advocate of evolutionary theory, entitles a chapter of his book *Darwin's Dangerous Idea*: "Sociobiology: Good and Bad, Good and Evil," and he is critical of what he calls "greedy reductionism," even in the work of E. O. Wilson, his fellow traveler. (Dennett, the philosopher, reminds us of "the genetic fallacy," which is the fallacy of reducing the meaning and the value of an idea to its origins.) He criticizes all simplistic extrapolations of ethics from nature, that is, *ought* from *is*. His particular targets are B. F. Skinner and Wilson himself, who proceed from different assumptions about nature. Though any rational ethical system necessarily suffers the constraints of human nature, ethics cannot and should not be determined by Nature. Does sociobiology tell us what we don't already know through other means that is relevant to ethics? In his concluding chapters on ethics, Dennett in a non-reductionist spirit is careful not to propose a Darwinian ethical system. A philosopher in the analytic tradition, he chooses rather to expose deficiencies in the rival theories of Mill's utilitarianism and Kant's *a priorism*. His book on Darwin's dangerous idea concludes with a benign vision of a democratic society, affirming not only biological diversity (the title of a subsection of a chapter) but cultural and religious diversity as well. Unlike the militant atheist Richard Dawkins, he acknowledges the gifts that religion has given to humanity as well as the theocratic dangers it poses. (He had not yet geared himself up for "breaking the spell.") So the concluding section in which he warns the reader that Darwinism can be a universal corrosive acid is something of a puzzle. "Corrosive acid" suggests the greedy reductionism that he is at pains to separate from what is essential to Darwinism. But if it is corrosive, what does it mean for him to defend the idea by asserting that it "preserve[s] and explains[s] the values we cherish"? (521). This reader at least finds nothing in this interesting book (interesting in what it tells us about the debates between Darwinists) that provides evidence for the view. Despite a valiant effort to turn the social and cultural implications of Darwinism into a benign humanism, Dennett in fact does betray the danger in Darwin's idea when it hubristically attempts to extend its domain beyond its areas of competence.

Dennett's book brought into focus a feature of the advocacy of Darwinian evolutionary theory that I perceived only dimly before reading it. In characterizing the debates within the community of biologists, he refers to the mainstream view as neo-Darwinism or the orthodox view. He has the unfortunate habit of speaking of the theory in a religious idiom. When Dennett takes on Stephen J. Gould, for instance, on the matter of gradualism vs. punctuated equilibrium, the scientific quarrel is inflected by the insinuation that Gould is heterodox vis-à-vis the Darwinian orthodoxy. Dennett even suggests that Gould's heresy may take him outside the Darwinian framework. To be sure, he makes rational arguments for the orthodoxy, but his language suggests an unscientific mindset of reflexive defensiveness to challenges to the orthodoxy. A symptom of this mindset is the constant invoking of Darwin. Science ideally does not have authority figures. Arguments in science are settled by an appeal to experiment and evidence, not by reference to an authority figure or to what Michel Foucault in "What is an Author?" calls a founder of discursivity such as Marx or Freud: "One defines a proposition's theoretical validity in relation to the work of the founders—while, in the case of Galileo or Newton, it is in relation to what physics or cosmology *is* (in its intrinsic structure and 'normativity') that one affirms the validity of any proposition that those men may have put forth. To phrase it very schematically: the voice of initiators of discursivity is not situated in the space that science defines" (116). No physicist invokes the authority of Newton or Einstein to settle a scientific quarrel. To do so would be reenacting the scholastic habit of invoking the theological authority of Augustine or Aquinas. This is not to deny Darwin's stature as a great scientist. He belongs in the pantheon of Galileo and Newton. But it does no service to the science he inaugurated for his followers in their debates to display something of a scholastic anxiety about whether the bounds of Darwinian orthodoxy have been transgressed.

The history of science has been a history of heresy. Since "evolutionary explanations are inescapably historical narratives" (Dennett, *Dangerous Idea*, 315), their scientific status is always contestable. Consider the following sentence: "The power of the theory of natural selection is not the power to prove exactly how (pre) history was, but only how it could have been, given what we know about how things are" (319). "Could have been" is an opening to alternative explanations. Dennett challenges the reader to describe some other way than natural selection to account for phenomena. Of course, for others to come up empty-handed doesn't necessarily establish the truth of his way. But he is right that Darwinism is the best we have (and it is a very good best), *pace* the creationists and the advocates of intelligent design. It is the best we have in accounting for the emergence of species. Where it overreaches is in its attempt to account for the achievements of the human species or when it tries to extrapolate an ethical system from the evolutionary process.

4

Is History a Science?

This is an old, but persistent question. In *Guns, Steel and Germs*, Jared Diamond, the distinguished physiologist and geographer, chastises historians for their failure to think of their profession as a science, the consequence of which is that they "receive little training in acknowledged sciences and their methodologies." It follows from this deficiency in training that much history writing is nothing more than "a mass of details…" or "just one more damn fact after another," or "more or less bunk" (421). (The last phrase is, of course, an unfortunate echo of Henry Ford's notorious dismissal of history.) Scientific history is the search for "ultimate explanation," in the instance of *Guns, Steel and Germs*, of why certain societies triumphed at the expense of others (European at the expense of African and Asian). His explanation, offered as a result of empirical investigation, lies in geography, that is, in the relative fertility of the land and the availability of animals that could be domesticated. The secret of success is food production, which frees people for a variety of other activities, for example the crafts.

Diamond is a geographical determinist. In his long, thirteen-thousand-year view of history, culture, ideology, personal agency, and luck play negligible roles. Impressive as he is as an empirical historian, he offers very little in the way of guidance to an understanding of the proximate as opposed to ultimate causes that make up the smaller, though by no means insignificant, time frames of human history, for instance, the French Revolution or the Russian Revolution and its aftermath. Indeed, even Diamond's large-scale vision does not exclude exceptions to his geographical determinism, which he simply glosses over. Consider, for example, his attempt to explain why China "lost its huge early lead to Europe," given its "undoubted advantages: a rise of food production nearly as early as in the Fertile Crescent; ecological diversity from North to South China and from the coast to the Tibetan plateau, giving rise to a diverse set of crops, animals and technology." Diamond explains it as "a typical aberration of local politics that could happen anywhere in the world: a power struggle between two factions at the Chinese court (the eunuchs and their opponents)" (411). The eunuchs favored "the sending and captaining of fleets," their opponents, prevailing in the power struggle "dismantled the shipyards... and forbade oceangoing shipping" (412). The first thing to remark is that politics not geography is the decisive factor here. "Typical aberration" is an oxymoron. If the factional struggle were an aberration, that is something exceptional, what would make it typical? Diamond neither raises nor addresses the question. One can only speculate that exceptional or aberrant actions that may proceed from human agency are an embarrassment to the scientific enterprise, which seeks out what is recurrent, inevitable or probable. The effect of "typical" is to neutralize aberration, which cannot be explained scientifically.

In speaking of non-scientific history as bunk, Diamond brings to mind those who are surely his adversaries, the radical skeptics who will settle for nothing less than absolute truth. Since the whole (or absolute) truth is an impossibility, the very idea of truth and objectivity is under suspicion. Diamond too has the all or nothing mindset, but unlike the radical skeptics he believes that he has it all. I begin with Diamond not because he has much to contribute to a theoretical discussion or debate about the scientific status of history writing, but because he represents what I would call an insurgent scientism that we find especially in contemporary Darwinism. Though Diamond views history mainly through the lens of geography and not genetics, he and Wilson have a declared affinity for each other in their ambition to place disciplines on a scientific basis. (Diamond is not averse to biological explanation as in his account of the way infectious diseases jumped from domesticated animals to human beings.) For Diamond there is no longer a question: history *is* or *should be* a science.

The title of E. H. Carr's provocative book on the subject, in contrast, takes an interrogative form: *What is History?* (1961). He defends history as a scientific enterprise by responding, unlike Diamond, to objections in a reasoned, if not, in my view, ultimately convincing manner. In establishing the scientific credentials of the historian, he gives the term science the widest latitude of meaning, noting that "in every other European language, the equivalent word to 'science' includes history without hesitation" (70). But is the equivalence exact? The German word *wissenschaft*, for instance, is a more comprehensive and looser term that may not in the English sense qualify as science. The hesitation about the use of "science" in the characterization of a discipline represents an English fastidiousness pointing to differences in the standards used for determining objectivity between historical research and scientific experimentation. In any case, the "term" science does not have the unambiguous precision of a value in a mathematical equation or of an electron or of a gene. One cannot simply assume that it is a defined enterprise waiting to be taken up by reluctant historians. What then are the specific grounds for viewing history as a science, or conversely what are the objections to such a view?

Carr addresses the following objections. "(1) history deals exclusively with the unique, science with the general; (2) history teaches no lessons; (3) history is unable to predict; (4) history is necessarily subjective, since man is observing himself; (5) history, unlike science, involves issues of religion and morality" (78). As to the first objection, Carr has in mind the view, originally expressed by Aristotle, that the province of history is the particular, not as with philosophy, the universal. It is a view that has a long history, to be found, for instance, in Leibnitz, who, as the French historian Marc Bloch tells us, when turning from abstract speculation on mathematics and theodicy to history, experienced "the thrill of learning singular things" (*The Historian's Craft*, 8). Carr challenges the view that history deals exclusively with the unique since "the very use of language commits the historian, like the scientist to generalization." And he points out that historians constantly make generalizations, for instance about "the causes of war" or the "habit of rulers of [a] period to liquidate potential rivals to their throne" (80). He is right in his challenge, though he might have allowed that historians are not obligated to rise from the unique and the singular to generalization. But he begs the question by not taking up the more serious objection that history does not produce laws in the scientific sense. Generalizations do not necessarily rise to the condition of scientific law, which involves predictability. In *The Poverty of Historicism*, Popper makes a useful distinction between laws and trends, which are not inexorable. Trends are susceptible of alteration by circumstance. Laws need to be revised if they are falsified by events. History books are filled with generalizations that may be illuminating. They are not, however, to be taken as laws.

The idea that we can learn from the past assumes that the past or aspects of it repeat themselves in the present. You don't have to be a conservative to believe this to be the case. But there is also historical change in which the past does not repeat itself. If this were not so, there would be no progress. Even those with a conservative view of history warn about the unintended consequences of actions and events. Historical generalizations or laws must account not only for recurrent patterns in history like that of the rise and fall of empires, but also for historical change which produces the unique or unprecedented events like a new form of social organization or a new economic system. I can think of few things as difficult, if not as impossible, as the formulation of historical laws that would cover the future as well as the past. Should such an attempt succeed, it would repeal the "law" of unintended consequences.

To the objection that history teaches no lessons, Carr answers that historians do draw lessons from the past to illuminate the present as well as lessons from the present to illuminate the past: "The function of history is to promote understanding of both past and present through the interrelation between them." One example: "It [is] dangerous, when redrawing the map of Europe, to neglect the principle of self-determination" (85), a lesson, one might add, that has never been absorbed by empires. If history repeats itself, as physical events following the laws of nature do, what does it mean to learn a lesson from history? Events will simply repeat themselves, whatever we learn from history. Or if history does not repeat itself, what are the lessons we learn from the historical past and how can we speak of historical laws? Yes, there are lessons we may learn from the past, such as the lesson that was not drawn from the British experience in the Middle East in the early twentieth century by the Bush administration when it invaded Iraq. But should we call this lesson scientific?

Whether historians can predict the future depends on what we mean by prediction. Carr views predictions as "statements of tendency, statements of what will happen, other things being equal, or in laboratory conditions. They do not claim to predict what will happen in concrete cases." Historians, Carr suggests, provide "general guides to future action, which though not specific predictions are both valid and useful" (87). Historians, of course, cannot test their predictions in a laboratory. Carr does concede that given "the complex rational entity" that is a human being, "the study of his behavior may well involve differences in kind from those confronting the physical scientist" (89). (He might have qualified "entity" with "irrational" as well as "rational.") He also concedes the subjectivity of history ("since man is observing himself"), and argues for "drawing a line of demarcation between the social sciences and the physical sciences." Since the subjectivity of history means the subjectivity of the historian, he draws support for eroding the line of demarcation by invoking the principle of uncertainty or indeterminacy

Is History a Science?   69

in physics in which "all measurements are subject to inherent variations due to the impossibility of establishing a constant relation between the 'observer' and the object under observation" (92). What Carr leaves out, however, is the subjectivity of the object under consideration, the men and women in history. There is nothing comparable to the difficulty of access to the intentions and motives of historical actors in the physical sciences. I say this in the knowledge that, as Popper has pointed out, "it is undoubtedly true that we have more direct knowledge of the 'inside of the human atom' than we have of physical atoms" (*The Poverty of Historicism*, 138). To be sure, quarks, neutrinos, et al. are inaccessible to the senses, but you can gain knowledgeable access through theory, mathematics, and experiment. And when you do, you will have discovered unambiguous truth, whereas the interior of "the human atom" is fraught with irreducible ambiguity and uncertainty. Natural scientists may concern themselves with the behavior of subjectivity-free eclipses, floods, volcanic eruptions and their history, whereas social and cultural historians have as their subject human experience, which, of course, includes the experience of natural events. However scrupulous and fastidious a historian of human life may be in recording the facts, his representations are filled with the often impenetrable subjectivities as well as the actions of persons. I suppose it is possible for historians to exclude systematically all reference to motive and intention and to confine themselves to actions and their tangible effects. But such a decision would produce an impoverished history.

Finally, Carr takes up the question of history's involvement with "issues of morality and religion." Historians, including those who do not regard the writing of history as a scientific discipline, would agree with Carr that the historian must not allow religious explanations (such as divine intervention) to enter into a view of events, whatever his religious beliefs may be. (We may assume, however, that explanations involving religion, such as the intimate relation between Calvinist Protestantism and the rise of capitalism are allowed.) Nor is the historian "required to pass moral judgments on the private life of the characters of his story" (96).

It is generally the case that those who subscribe to a scientific view of history are reluctant to grant much space to the actions of individuals, perhaps because it is easier to quantify social forces than individual actions. (It is impossible to quantify individual intentions and motives.) Carr avoids the problem by rejecting the idea that there is an "antithesis between the individual and society" (69). Following Hegel, Marx's teacher, Carr views the great man (the individual who is said to count in history) as "one who could put into words the will of his age, tell its age what its will is, and accomplish it" (68). Carr's formulation is more pedestrian: "The great man [is] an outstanding individual who is at once a product and an agent of the historical process, at once the representative and the creator of social forces which change the shape of the world and the thought of men." It should strike the reader as strange that a historian whose aspirations are scientific would have recourse to something so mystical as "the will of the age," as if it were something singular and discernible prior to its actualization by the great man. How would Carr or anyone else know whether Napoleon or Lenin or Hitler had actualized a "will" that no one except the great man could perceive before its actualization? And does history have only one will at any of its moments or periods? Moreover, the identification of the individual with the great man excludes those individuals and groups who may never achieve power, but who nevertheless find themselves at odds with social authority, and in their opposition to or alienation from society may affect the course of its history. I have in mind philosophers, literary artists, scientists, as well as social and political groups. The individual may or may not be antithetical to society. Carr's insistence that there is no antithesis between them reflects a certain "scientific" view of history that we find, for instance, in Marxism.

But let us for the moment stay with the great individual who has done so much to shape and to actualize "the will of the age." He has no moral interest for Carr. "What profit does anyone find today in denouncing the sins of Charlemagne or of Napoleon? Let us therefore reject the notion of the hanging judge, and turn to the more difficult but more profitable question of passing moral judgments not on individuals, but on events, institutions or policies of the past" (100). One may indeed pass judgments on institutions and policies (I am not sure what it means to pass judgments on events), but institutions and policies are the responsibility of individuals or groups of individuals. It may not be profitable to denounce Charlemagne or Napoleon (figures from an incorrigible past), but we should surely want to bring to justice leaders of nations responsible for crimes against humanity. Carr here betrays a bias that views history as the exclusive product of impersonal social forces despite his concession that individuals may have a role in actualizing them. The test of the significance of individual agency is in contemplating what would have happened if Napoleon or Lenin or Hitler had not come into existence, or if, as Philip Roth has imagined in a recent novel, Lindbergh, not Roosevelt, had been elected president in 1940. In fact, Carr implicitly concedes that the history of Russia would have been different if Lenin had not been present to lead the Bolshevik revolution, when in listing its many causes he mentions "the fact that Lenin knew his own mind and nobody on the other side did" (116).

Like scientists, historians look to causes. The discovery of a chain of causation leading up to, say, the American War for Independence or the French Revolution may lead to claims that what unfolded was "inevitable." Inevitability is a term in the discourse of scientific determinism. In the view of philosophical liberals like Karl Popper and Isaiah Berlin, determinism when applied to social and political matters has a history of association with despotism. Determinism provides no space for freedom of the will. Carr has little patience with this view and dismisses it as a "red herring" (119). He sees no harm in determinism, which he considers a commonsensical view of human life. He defines it "as the belief that everything that happens has a cause or causes and could not have happened differently unless something in the cause or causes had also been different" (121-22). But determinism must mean something more than the commonsensical doctrine that events have causes. It must also mean that events can be shown to obey or follow laws. Contingent and random causes (for instance, a car driven by a drunken driver hits and kills a cautious pedestrian) cannot be subsumed by laws, except by a trivial one such as drunk drivers threaten the lives of innocent pedestrians. What should we make of the decision of a leader who has to decide on a course of action in a crisis? What if the leader is indecisive and the choice he makes is unpredictable. He may have chosen a course of action different from the one he had originally decided upon. In which case, determinism is a highly problematic concept.

Establishing a lawful chain of causation is not a simple and uncontroversial matter. Consider Carr's view of the multiplicity of causes of the Bolshevik revolution: "Russia's successive military defeats, the collapse of the Russian economy under pressure of war, the failure of the Tsarist government to solve the agrarian problem, the concentration of an impoverished and exploited proletariat in the factories of Petrograd, the fact that Lenin knew his own mind and nobody else did—in short, a random jumble of economic, political, ideological and personal causes, of long term and short term causes" (116), It is to Carr's credit that he is not guilty of reductionism, the sin of other scientifically inclined historians, which Marc Bloch memorably indicts: "The sheer complexity of so much of the 'matter' that [the historian] has to encompass and its insusceptibility to general laws require a sense and sensibility quite different from those of a natural scientist." Bloch is wary of "*homo religious, homo oeconomicus, homo politicus* and all that rigmarole of Latinized men...phantoms which are convenient providing they do not become nuisances." They become nuisances when "the man of flesh and bone...the only real thing is forgotten" (51).

On the matter of determinism, however, Carr remains vulnerable. "A random jumble" would seem to rule out a chain of causation. What weight should we give to each of these causes, and how are we to evaluate the interaction among them? I think we can rule out scientific quantification. And how does Carr know that only Lenin knew his own mind or that knowing one's mind necessarily leads to a successful revolution? He is willing to substitute "a high degree of probability" for inevitability, but he betrays his inevitabilist bias when he mocks a topic offered by a lecturer, "Was the Russian Revolution inevitable?" (126) as a futile exercise in trying to imagine a sequence of causes that did not exist. He has no patience for counterfactual reasoning, which may be necessary in achieving an understanding of what actually occurred. He would have benefited from Fritz Ringer's succinct summary of Max Weber's case for counterfactual reasoning: "To identify significant singular causal relationships at all...we must inquire into the degree to which a particular cause 'favored' a given effect. But this in turn requires us hypothetically to 'compare' the result that actually followed with alternate possibilities. The historian cannot avoid reasoning about historical events that *did not occur*, in order to identify the significant causes of *what did occur*" (Ringer 83). Counterfactual reasoning should be a tool, indeed a scientific tool, in the kit of every historian. Unless we have the imaginative freedom to consider alternative scenarios and outcomes, the only lesson we can draw from history (Carr, we should recall, believes that lessons can be drawn) is that we have to accept stoically whatever history deterministically provides. Carr simply dismisses the anxious concern expressed by Popper and Berlin about the despotic political use to which the doctrine of determinism has been put. We have already had an excess of experience of regimes that invoke "the laws of society" to justify political repression.

A "scientific" history with its Enlightenment bias would necessarily be "progressive," first in the historiographical sense and then in the political. Historiography, Carr asserts, is progressive in that knowledge is cumulative. With the widening and deepening of knowledge present understanding supersedes past understanding, future understanding will supersede present understanding. While it is true that present knowledge (in the sense of information) is greater than past knowledge, it doesn't necessarily follow that present understanding supersedes past understanding and that future understanding will supersede that of the present. Have contemporary historians superseded Thucydides or Gibbon or Hume or Parkman in the quality of understanding? As for political perspective, the historian with a scientific or rationalist bias tends to favor reason over tradition and "hope for the future" over "reverence for the past." I am quoting from a review essay on Edmund Burke by J. H. Plumb who, while paying lip service to Burke as "a great philosopher," cannot conceal his contempt for his conservatism. "The rationalization of prejudice, the sanctification of the status quo, the attribution of historical inevitability and Divine Providence to inequality and human suffering certainly acquired its most persuasive apologist in Burke and so, perhaps it is not surprising that he is rapidly becoming a cult" (*In the Light of History*, 95-96). Plumb's attitude toward Burke is not fortuitous; it is consistent with a commitment to the ideals of Science and Reason. It is hard to resist a rejoinder: you don't have to be a conservative to be impressed with the fact that human beings have always suffered and that the probability is high that humanity will continue to suffer in the future. You may *reasonably* hold the view that inequality is a chronic condition that should be ameliorated by reform, but that the utopian attempt to eliminate it entirely may make things much worse. Nor does one need to belong to the cult of Burke to recognize that his view of tradition allows for change as well as stability. He did not sanctify the status quo when he supported the American colonies in the War for Independence. The belief in Science and Reason does not necessarily guarantee objectivity.

The idea of progress haunts the quarrel between science and the humanities. Built into the project of science is the belief that it is making progress in revealing the truth of our moral as well as physical existence. In the early phase of the Enlightenment the progress of knowledge was intertwined with moral progress. The improvement of the human lot went hand in hand with scientific discovery. (The word evolution itself carries progressive connotations, though its exponents resist the identification of progress with evolution.) If only scientists ruled the roost, our lives would be all the better for it. Skepticism about moral progress is at the core of the humanist vision not because humanists do not desire progress, but because they do not see empirical evidence for *moral* progress in post-Enlightenment history. So humanists may value the idea of progress while believing that it represents only in part the actual history of civilization. Science and technology have brought us improvements in medicine, in communications, in the amenities of life, it has also increased to an extraordinary degree the capacity of human beings to destroy one another. This is the familiar seesaw paradox of scientific progress, which does not always communicate to the thinking of the champions of science about their enterprise. You can hold to an ideal of progress and at the same time believe that the movement of civilization is not always a forward one in a positive sense, or even that science may not in all cases be a vehicle of progress. In this sense, the idea of progress is less an historical fact than an ideal to measure particular achievements. Moreover, progress entails losses as well as gains. The backward looking of literary humanists, which C. P. Snow complained about, is at its most serious not nostalgia, but rather an effort to preserve what is of value in the past against the "progressive" corrosions of history.

* * *

Those of us who live in the humanities or the social sciences may be under the illusion that postmodern relativism has dethroned science. The historians Joyce Appleby, Lynn Hunt, and Margaret Jacob write of "the dethroning of science as the source and model for what may be deemed true" (*Telling the Truth about History*, 13). Objective knowledge, it would seem from what radical skeptics say and write, is no longer possible. They speak of an insurmountable gap between the language we use and the real world. Our report of the world at best refracts rather than reflects what we observe. Personal bias and ideology "contaminate" observation. The above statements are not equivalent, but they share an animus against the possibility of objectivity. Why then turn one's critical attention to science or scientism, the great exponent of objective knowledge? Is it a case of beating a dead horse? First, let me emphasize again that my target is not science but scientism, an *ideology* that promotes the view that science, or as it may turn out a debased version of it, should define all the disciplines. In its hubris, it is the mirror image of the radical skepticism that denies the possibility of objective knowledge, even in the sciences.

I speak of the dethroning of science as an illusion, because science continues to exert and validate its power in the world through extraordinary discoveries (DNA, the human genome) and technological achievements (in medicine, in physics etc.). The power of radical skepticism is confined to a section of the academy and to its amplification by the media, which exploit it for its entertainment value. (I don't underestimate its effect as I have written extensively and critically on the subject.) In any event, practicing scientists, (not all) theorists and historians of science tend either to ignore or to dismiss skepticism directed to their work in their justifiable confidence in their achievements. Which is not to say that the ignoring of skepticism is justifiable. It is in the interest of all the disciplines to resist hubris, which means cultivating a healthy, not radical, skepticism about their own enterprises. This is especially case when a discipline attempts to aggrandize other disciplines in the belief that what works for it will necessarily work for others.

Certainly, we would or should want to preserve the idea of objective knowledge for historical study. A historian with postmodernist or post-structuralist sympathies like Hayden White might not find it a cause for concern, but for most historians the survival of the discipline would be at stake if, without a belief in objectivity, it were to become simply a species of fiction. According to Appleby, Hunt, and Jacob, "history is crucially distinguished from fiction by curiosity about what actually happened in the past" (259). The crucial word here is "actually," for fiction too may be curious about the past, may indeed in certain instances provide greater illumination of the past than one might find in a comparable work of history. What distinguishes history from fiction are the constraints it imposes upon itself to accumulate and not violate the facts, even as it goes beyond them in interpretation and speculation.

In preserving the idea of objective knowledge in historical inquiry, we need to distinguish its conditions from those that obtain in the natural sciences. I anticipate the rejoinder of the advocates of scientific history: "Of course, we are willing to draw a line between the natural sciences and history." We have seen the willingness in Carr. Where, then, is the quarrel? I would draw the line between history and science in the following manner. There are scientific tools in the tool kit of the historian, statistical analysis for instance, which he is free to employ. But the task and disposition of the historian is to tell a story of events that occurred (the objective element) that may or may not achieve the status of generalization and will certainly not achieve that of law. The subjective, interpretative character of historical representation and the elusive, sometimes impenetrable, subjectivity of the objects, or more accurately, the subjects being studied disqualify the discipline of history as a science. History is susceptible to constant revision of perspective. The trajectory of revision is not linear and progressive as it tends to be in science. What influences the revision of historical perspective is history itself, that is, the changing historical situations in which the historian finds himself or herself. As with science, new evidence may come into view to alter historical interpretation and understanding, but unlike with science in its ideal expression, history or history writing is inflected by the political and moral biases of the historian. In interpreting historical characters and events, the scrupulous historian tries to be faithful to the facts and as disinterested as possible in his conception or construction of the interpretative pattern that will illuminate the facts. By disinterested, I mean the effort to perceive a pattern that may not confirm an a priori bias or interest of the historian, indeed, may even go against it. Though we cannot escape our subjectivity, we can mitigate or minimize its distortions through the *effort* of disinterestedness. The historian knows or should know that he approaches his subject with a point of view and that his capacity for objectivity depends upon a critical awareness of its limitations. He also knows that the period and events that he or she has selected for study can and will always be subject to alternative interpretations. Unlike science, history does not aim at singular explanation that trumps all other explanations. A classic work of history may be considered definitive only for a time. There is no final theory or interpretation possible in historical inquiry. Unlike any of the sciences, history does not aspire to unambiguous mathematical representation. Whatever truth it discovers is to be found in the vernacular in which inventiveness and elegance of expression, as evidenced by the examples of Thucydides and Gibbon, matter.

What Appleby, Hunt, and Jacob call "heroic science" was born in the Enlightenment, its adversary the formidable authority of religious dogma, in its eyes institutionalized superstition. Valuable in its own time, it now suffers from hubris. But the legacy of objectivity that it has bequeathed needs to be defended if the boundaries between empirical truth and fiction are not to be eroded.

Postscript: It should be evident that I am concerned here with human, not natural history. Evolutionary theory is a theory of natural history that claims to be a science. In endorsing this claim, we will have to make allowances for the fact that the knowledge it yields is retrospective and not prospective. Evolutionary theorists have not formulated laws that can predict our evolutionary future. Karl Popper refers to "a brilliant scientific hypothesis concerning the history of the various species of animals and plants on earth" (106), but he also asserts controversially that "the evolutionary hypothesis is not a universal law of nature" (107). This is not to say that he doesn't acknowledge that "modern Darwinism [is] the most successful explanation of the relevant facts" (106). It is interesting to see Popper worrying about the scientific status of evolutionary theory at the same time that he recognizes its explanatory powers. In the footnote in which he pays tribute to Darwinism, he begins by confessing to "the feeling of being intimidated by the tendency of evolutionists to suspect anyone of obscurantism who does not share their emotional attitude toward evolution as a 'daring and revolutionary challenge to traditional thought.'" Evolutionary theory has made considerable advances since the time of Popper, but his questioning, even ambivalent attitude toward it when its reach exceeds its grasp is salutary.

5
Condescending to Science

Misadventure is a two way street. Humanists too have overreached in their quarrel with Darwinism. In his recent book *Darwinism and Its Discontents* (2006), Michael Ruse makes brief reference to the discontents, not all humanists. Here are some samples:

> It is not to be expected of Darwin that he should have been troubled by thoughts of fallibility, relativity or indeterminacy; but only that he should observe the standards of his own time. And it was by those standards that he was in arrears. Nineteenth-century science was sufficiently aware of the desirability of precision and standardization to make Darwin's tool chest seem distinctly unprofessional. In this, as in other respects, he gives the appearance of an amateur, an amateur even for his own day (the historian Gertrude Himmelfarb in 1959, quoted in Ruse, 5).

> The believer in God, unlike her naturalistic counterpart, is free to look at the evidence for the Grand Evolutionary Scheme, and follow it where it leads, rejecting that scheme if the evidence is insufficient. She has a freedom not available to the naturalist. The latter accepts the Grand Evolutionary Scheme because from a naturalistic point of view this scheme is the only visible answer to the question *What is the explanation of the presence of all these marvelously multifarious forms of life?* The Christian, on the other hand, knows that the creation is the Lord's; and she isn't blinkered by a priori dogmas as to how the Lord must have accomplished it. Perhaps it was by broadly evolutionary means, but then again perhaps not. At the moment, 'perhaps not' seems the better answer (the philosopher of religion Alvin Plantinga in 1998, Quoted in Ruse, 25).

> Evolution: The Fossils Say No! (Young-Earth Creationist Duane T. Gish in 1973, Quoted in Ruse, 72)

> The Synthetic Theory of Evolution is Dead. (the paleontologist, Stephen J. Gould in 1980. Quoted in Ruse 103)

> Darwin's theory has no more relevance for philosophy than any other hypothesis in natural science (the philosopher Ludwig Wittgenstein in 1923. Quoted in Ruse, 236)

[And finally] There is a hideous fatalism about it, a ghastly and damnable reduction of beauty and intelligence, of strength and purpose, of honor and aspiration, to such casually picturesque changes as an avalanche may make in a mountain landscape, or a railway accident in a human figure. To call this Natural Selection is a blasphemy, possible to many for whom Nature is nothing but a casual aggregation of inert and dead matter, but eternally impossible to the spirits and souls of the righteous. (G. B. Shaw in 1898. quoted in Ruse, 258).

Each of these quotations is at once an epigraph and a foil for successive chapters in which Ruse makes out his case for evolutionary theory. Each of the objections to Darwinism proceeds from different grounds, depending on the professional affiliation of the critic. Himmelfarb finds Darwin to be second-rate, if not incompetent, according to the scientific standards of his time as she a historian understands them. (I will say more about Himmelfarb below.) Plantinga, a professor of religion and a man of faith, perversely bases his faith on empirical knowledge against the "a priori dogmas" of naturalism. Gish is a creationist who claims to know the fossil record. Gould, the biologist, is the surprise for those who have not followed the controversies within the field of evolutionary theory. He is an evolutionist for whom natural selection has a very diminished role to play and who is hostile to the genetic bias of the neo-Darwinian synthesis. Regardless of the side that one might take in Gould's quarrel with neo-Darwinism, the fact is that the Synthesis is not dead. Wittgenstein is not dismissive of Darwinism, but, as he says, he fails to see its relevance to philosophy. And Shaw, the dramatist is a vitalist in the company of Samuel Butler and Henri Bergson, contemptuous of materialism. All these quotations convey the range of discontent with Darwinism, much of which is unearned, proceeding as it does from ignorance or insufficient knowledge. Some of it reflects defensiveness, a sense of threat to time-honored assumptions about the meaning of human existence.

In 1941, Jacques Barzun published *Darwin, Marx and Wagner* in which he took each to task for his materialist approach to science, society, and the arts respectively. In Barzun's account, materialism becomes the dominant way of looking at the world in the nineteenth century. His complaint against Darwin focuses on natural selection, which is purposeless. His sympathies are with Darwin's critic, Samuel Butler, who saw in the evolutionary process the workings of mind. "Starting from the truism that the organism is a living thing and not a machine, Butler asserted that its characteristic feature is that it has 'an interest': it wants to do certain things and not to do others. Mental here refers not to Intellect, but to consciousness, however low and limited.... Mind, feelings, ethics, art—all these things once again become real, instead of being the dreams of automata, accompanying the physico-chemical changes called digestion, respiration, reproduction and death. It was still possible to look at these functions in their physico-chemical aspects, but to do so did not exhaust the meanings inherent in them" (108-09). Butler, Barzun explains, "wholly disclaimed for his theory any scientific warrant or originality" (110). Rather he proposed a vitalist alternative to both materialistic science and theological creationism. The line from Butler to Barzun goes through Bergson, and Shaw, who preached vitalism against Darwinian scientism. (Nietzsche also has a place on this line.) If vitalism has no scientific warrant, then in what sense is it an alternative to Darwinism? Wouldn't it be more accurate to say that to the extent that it does not intrude upon the scientific ground of Darwinism, vitalism represents a complementary vision of life, which finds a congenial home in imaginative literature and philosophy. To the extent that vitalism presents itself as a rival to Darwinism, it is guilty of appropriating territory that does not belong to it.

There is an equivocation in the Barzun/Butler view between an argument for an organicist or vitalist approach and a concession to a materialist approach to physico-chemical aspects. What Barzun does not say but implies by his argument is that for anyone who takes the materialist claims of science seriously (which means taking science seriously) and yet wants to avoid a totalizing materialism dualism is unavoidable. The Barzun/Butler view does not amount to Intelligent Design as its contemporary advocates present it. The mental activity is consciousness "low and limited," and God does not enter the picture. It is, however, a view that persists in certain mostly non-scientific quarters that the case for natural selection as the vehicle for evolution has not been convincingly made.

In *A Stroll with William James* (1983), Barzun softens his critique of Darwinism without relinquishing it. He cautions against "the unfortunate resurgence of a 'Creationist' movement of opinion [which] adds to the confusion of rekindling the scientists' fears and crusading ardors, especially since Bible believers have seized on the 'Big Bang theory' of neighboring physics to use as proof of creation." He then declares himself to be "an evolutionist, perhaps of a unique sort. I am not a Darwinist but until further notice I am a complete Darwinian: I believe in the grandfather [Erasmus Darwin] and the grandson and the holy spirit of their inquiries. On the one hand, the thought of innumerable separate creations by a Master bio-engineer defies probability and many facts. On the other, the story of protoplasmic blobs shooting off little changes in all directions until, by piling up useful accidents of shape they wind up as a profusion of complex creatures overlooks the presence of coordinate arrangements—of design" (213). Barzun's parti-pris, of course, doesn't have scientific authority, as he himself admits. He simply doesn't know enough (neither do I), but it does reflect what an intelligent humanist in good faith reading about the tangled controversies within the community of neo-Darwinism might reasonably conclude. (One wonders how Barzun would reconcile the Lamarckism of Erasmus with the Darwinism of his grandson.)

There is the public problem for Darwinists, who avoid controversy with advocates of Intelligent Design because of the difficulty of making the case to a non-scientific audience. For the non-scientist to be able *knowledgeably* to embrace a scientific argument, he would have to possess the data as well as methods of operation that come with the territory of being a scientist. As non-scientific humanists, we endorse natural selection (as we do quantum mechanics) as an act of faith in the absence of the knowledge that evolutionary biologists possess. (It should be noted that a perusal of the scientific literature on the subject of natural selection reveals disagreements among Darwinists not about its existence, but about the extent to which it determines the evolutionary process.)

In 1959, the hundredth anniversary of the publication of *The Origin of Species*, Gertrude Himmelfarb published *Darwin and the Darwinian Revolution*, in which she concludes that while "posing as a massive deduction from evidence, [natural selection] ends up as an ingenious argument from ignorance" (319). Himmelfarb does not offer an alternative explanation or theory of evolution. She is content simply to make a case against Darwin. What she does not provide is the evidence or the lack of evidence for natural selection. Indeed, it would be difficult to see how she could do that even if she had the scientific knowledge of paleontologists. (She points out that natural selection does not account for the origin of variations, which are selected. But this should hardly count as an objection. The science of genetics, the advent of which postdates Darwin's theory, does account for variations and complements natural selection.) Over the past decades a vast amount of data in the fossil record and in molecular biology has accumulated. Mainstream scientists are not only more convinced than ever that natural selection and adaptation are the principal, if not exclusive, explanations of evolution, they hold the view that Intelligent Design, let alone creationism, doesn't even deserve a hearing. No respectable Darwinist will even condescend to debate the matter. Himmelfarb's attack on Darwin and Darwinism, it would appear, has behind it an animus against science itself. "It is a common habit of the scientist not only to reduce the complicated and sophisticated to the simple and primitive but also to prefer simple and primitive. He will cheerfully contemplate the reduction of the most exalted belief to the mechanical or instinctive act, but will not tolerate a belief that, in all its complex reality, happens to conflict with his rational prejudices" (353). Her complaint is undiscriminating. One needs to distinguish between the uses of reductionism. Reductionism in physics, for instance, can provide necessary and fruitful explanations. She fails to appreciate (or acknowledge) the imaginative and complex effort that goes into arriving at a reduction such as $e=mc^2$. In conflating "the simple and the primitive," she ignores the extraordinary intellectual sophistication required in achieving the economical, not simple*minded*, result of a scientific law. There is, to be sure, a difference between simplicity in the arts and in the sciences. Something of that difference is conveyed in Wilson's sentence that I have already cited: "The love of complexity without *reductionism* makes art, the love of complexity with *reductionism* makes science." That doesn't say it all. The interpretation of art not only acknowledges the complexity, but in a sense increases it. This is not to say that there is or should be a conflict between the sensibilities. Each has its necessary function in the discipline in which it operates.

Barzun and Himmelfarb do not figure in contemporary controversies, but they anticipate what continues to be a certain discomfort among humanists with what might be called the reach of neo-Darwinism. In *Science and Poetry*, the philosopher Mary Midgley affirms the organic wholeness of the world not only against the mechanico-materialist reduction of it to discrete atoms, but also against its extension to the social world. What Midgley calls social atomism, the product of reductionism, cannot in her view provide an adequate account of emotions, passions, imagination or of social life. The poets Wordsworth and Keats become witnesses for her anti-materialist, anti-reductionist view. Midgley does not consider herself as hostile to science. She is, however, partial to the cause of poetry in its quarrel with scientific materialism, a partiality that sometimes distorts the target of criticism. For instance, she identifies scientific materialism with Dickens's gross caricature of utilitarianism in *Hard Times*. Gradgrind's insistence on "facts, facts, facts" does not begin to take into account the theoretical and analytical aspects of science. Nor does James Mill's utilitarian education of his son John Stuart exemplify a Gradgrindian education, whatever Dickens may have intended. If Mill failed to nurture the emotional side of his son's nature, he did cultivate his intellectual powers, which were world's apart from the mechanical rote of Gradgrind's pedagogy.

Like other humanistic anti-reductionists, Midgley wants to save science from what she regards as its worst tendencies. She has written a number of books in which she puts forth the organicist philosophy of Gaia (the name of the earth goddess). In *Science and Poetry*, her main targets are the neo Darwinists. Midgely writes: "By showing the earth as a whole rather than a loose heap of replaceable resources, it makes it possible for us to see it as vulnerable, capable of health or sickness, capable of real injury" (16). Midgley has her own version of consilience, a word, however, she does not employ. Like Wilson, she believes that all the disciplines and the arts should unite in the effort to save the planet. She is at one with him in her desire to preserve biodiversity and prevent ecological extinction. What is at issue between them is the question of the nature of that unity and wholeness. At times she seems to allow a possible role for reductionism when she concedes that it *is* a rational way, though not the "better" way:

> In short, reductionism is not the only rational way of dealing with differences of scale. There are much better ways of representing them. Different forms of order can co-exist at different levels, so scientists can use different ways of thinking about them without fighting, without insisting on reduction and without scandal. (6)

But elsewhere she is more severely dismissive. She holds with Brian Goodwin that "conflict only arises when there is confusion about what constitutes biological reality," and she assumes that the confusion and therefore conflict can be resolved if only we agree that reality is an *organic* unity. One might think that conflict would also be resolved if there was agreement that reductionism is the way to the truth. But, of course, this is not possible if the conflict is between science and poetry, for the philosophy of poetry in its romanticist version is that the world is an organic whole and therefore inimical to reductionism. Only its adoption by science under the aegis of Gaia (the earth goddess) will produce the harmony we all desire. In Midgley's view, the one-way reduction of wholes to parts in the natural and social sciences (organisms to genes, society to individuals) results in the impoverishment of our conception of wholeness, of the interdependence of our world. Some of its consequences are social atomism, determinism (i.e., the evaporation of the idea of freedom) and the reduction of mind and imagination to the physiology of the brain. (Midgley thinks in terms of two alternatives: either unity or harmony between science and humanities or conflict between them.

But disciplines may have different goals that may not be in harmony, but not necessarily in conflict.)

Social atomism is the extension to society of the scientific view that the material universe is reducible to the elements that compose it; where individuality, solitude and privacy are valued above everything else. Social atomism is inimical to community and mutuality, the idea that we belong not only to ourselves, but also to each other. It is antithetic to the conception of the world as an organic whole that needs to be cherished and protected. Reductionism not only promotes an anti-communal individualism, it also affirms determinism. Midgley cites Karl Popper who speaks of determinism as "*a daydream of omniscience*" (the emphasis is Midgley's) "which seemed to become more real with every advance of physics until it became an apparently inescapable nightmare" (26). By postulating a fixed order governed by a chain of causation, it flatters our vanity that we have understood and mastered the world. But it is an illusory mastery, for in its fatalism it effectively rules out freedom. Bad philosophy rather than science, determinism "is an assumption that goes infinitely beyond any possible evidence, one that is made for the sake of its useful consequences." Midgley's sentence is puzzling, because in admitting to the usefulness of determinism, albeit a limited one, she has undermined her claim that "it is an assumption that goes *infinitely* beyond any possible evidence" (my emphasis). I sympathize with her skepticism about determinism as it applies to human affairs, but she does not make a persuasive argument against determinism as an assumption in the physical sciences. Moreover, it is by no means clear that her own organicist vision of life escapes the nightmare of determinism. Her disparagement of individuality, solitude, and privacy seems incompatible with an affirmation of freedom. The idea of an organic society has a history of association with political and social repression.

Like her sociobiological adversaries, Midgley is a monist. Dualism is her bête noir. The social atomism, anathema to Gaia, would follow from scientific reductionism in the natural sciences only in a monistic conception of the world. Dualism, in contrast, would confine atomism to the scientific understanding of the physical world. It does not follow from the principle of causation in the *physical* world that human actions are absolutely determined and that there is no space for human freedom. It may be true that the scientific habit of mind has a bias in favor of deterministic explanations in the human as well as the physical world. But not all scientists have this bias. Dualism of the sort that the physicist Steven Weinberg advocates (see below) is in fact a caution against it.

Midgley affirms the "need somehow to consider mind and body not as distinct items but as complementary aspects of the whole person" (99), meaning, one may surmise, that they interact and contribute to the formation of our personality. By arguing that mind and body are complementary, Midgley means to avoid both idealism and materialism, an attractive prospect on the face of it. But it is not at all clear what complementarity, as Midgley conceives it, would mean for the scientific study of the brain. Dualism does not deny that mind and body interact, but it allows the study of the brain to proceed on materialist principles without reducing mind and imagination to the physiology of the brain. I don't see how dualism in one form or another is avoidable. Should we deny ourselves whatever knowledge we might gain about ourselves from the study of the (deterministic) physiology of our brain? I cannot imagine a sensible answer in the affirmative. Gaia in its romantic poetical appropriation of science is neither a solution to the problem of knowledge nor a guarantee of human freedom.

In keeping with her desire to unify the disciplines, she protests against what she characterizes as "a strange kind of apartheid in the teaching of literature." Her examples are Conrad's sea stories in which storms and other natural phenomena are treated as "scenery for the human dramas involved rather than as a central part of their subject matter," and the science fiction of H. G. Wells "and the whole vigorous tradition of science fiction," which "has been long cold-shouldered out of the literary syllabus," despite "Conrad's and James's admiration for Wells" (22). As readers of Conrad's fiction, we should expect keen observation and accuracy in his description of the behavior of a storm, though not necessarily a scientific understanding of it. And what exactly does Midgley mean by "natural phenomena" being "a central part" of the subject matter of the sea stories? The interest of a storm in Conrad's fiction is in its metaphorical, not meteorological aspect, which means the way it is appropriated by the human drama. Whether Wells deserves to be taken more seriously by literary scholars is a matter for literary, not scientific, judgment. Which is not to say that scientific subject matter has no place in literary works. From the nineteenth century to the present, Darwinism has been a theme of the literary imagination. Lydgate, a major character in *Middlemarch*, has as his great ambition the discovery of the "primitive tissue that is the basis of life," (an exercise in reductionism!) and *The Magic Mountain* would be drained of subject matter without its scientific interest. The question is not whether literature should take an interest in scientific matters, but whether there is a difference between a scientific and a literary perspective. I believe there is a difference, just as there is one between, let us a say, a philosophical disquisition on the nature of freedom and a novelistic treatment of the idea as it is embodied in the career of a character. It is the difference between Hegel and Dostoevsky. And it is a difference that does not preclude commerce between them; indeed, without the difference there can be no commerce.

6

In Defense of Dualism

Physicists may dream of a final theory of the physical universe, but they tend to resist the temptation to extend their ambition to human affairs. Biologists, by contrast, view all of life on a continuum. If so disposed (and not every biologist is so disposed), they see nothing in theory or practice to stop them from moving down along the continuum from the complexity of human life in all its manifestations to the simple beginnings from which complexity was generated. And they may feel no compunction in finding in the simplicity a full explanation for the complexity. What they fail to appreciate is that the road traveled to poetic imagination, for instance, from its beginnings in its cellular or molecular source in the brain brings us to a new and different place. The understanding of the meaning and power of an artistic or intellectual achievement is irreducible to biological explanation. In contrasting the physical sciences and the life sciences, I don't mean to suggest that professional differences fully account for the positions individual scientists take. There are life scientists who view complexity not simply as an increase in quantity, but rather as a qualitative transformation of its origins.

The physicist Steven Weinberg (a Nobel laureate) refers to himself as a dualist in the following way: "I think that an essential element needed in the birth of modern science was the creation of a gap between the world of physical science and the world of human culture." The gap was the creation of René Descartes, who provided modern science with its philosophical justification. Dualism makes those who strive for unified theories uneasy. Perhaps modern science at its inception needed to be liberated from the cultural domination of theology in the seventeenth century, but is that a reason not to aim for the unity of knowledge in the twenty-first century? One might expect that Weinberg, who dreams of a final theory which would give us "the ultimate laws of nature," would not be content with dualism, but apparently he is. The laws of material nature are not those of culture and society. He is implicitly at odds with sociobiologists who seek to bring all of human experience under the aegis of science. "Endless trouble," he remarks, "has been produced throughout history by the effort to draw moral or cultural lessons from discoveries of science" (*Facing Up*, 156). His example from the distant past is the justification of slavery on the basis of Aristotle's conception of nature.

He could have cited as well the Social Darwinism of Herbert Spencer, which invokes the survival of the fittest in the subhuman animal world (an accepted scientific view) to justify the ruthlessness of the capitalist marketplace. Sociobiology is the most recent instance of drawing moral or cultural lessons from the natural sciences, and it is still an open question about how much trouble it has produced or will produce. Apropos of Marx's claim "to be guided…by a science of history," Weinberg asserts that "of course, [there is] no science of history" (*FU*, 248). Marx, he argues as others have argued before him, is a utopian in denial. (Both Marx and Engels disdained utopian thinking.) "If Marx had been an honest utopian, and recognized his responsibility to describe the society he wanted to bring into being, it might have been clearer from the beginning that the effort would end in tyranny." By invoking the authority of science, Marxism conceived the course of history as an inevitability in which human agency is transformed into a mere instrument of that inevitability. Here is a clear instance of the freedom-depriving effect of scientism (not science!). The objection to these scientific "lessons" is that they are at once false and morally destructive in their applications.

Where human affairs are concerned, I believe that Weinberg would subscribe to the view held by Isaiah Berlin. Berlin dismisses the antiscientific animus of those who insist that "natural science and the technology based on it somehow distort our vision." He speaks of this view as "absurd nostalgic delusion." But he is equally dismissive of the view that "everything" can be "grasped by the sciences." He invokes the vision of human history to be found in Tolstoy, "who taught us long ago, [that] particles are too minute, too heterogeneous, succeed each other too rapidly, occur in combinations of too great a complexity, are too much part and parcel of what we are and do, to be capable of submitting to the required degree of abstraction, that minimum of generalization and formalization—idealization—which any science must exact" (48-49). But, one might object, science and scientific theory are still in their infancy with respect to human affairs. Isn't it possible that in the indefinite future, the sciences will achieve the necessary sophistication to master the human world? In response to this kind of question, Berlin is care-

remarks, no scientific theory can substitute "for a perceptual gift, for a capacity for taking in the total pattern of a human situation, of the way in which things hang together—a talent to which, the finer, the more uncannily acute it is, the power of abstraction and analysis seems alien, if not positively hostile" (50).

* * *

A salutary, if not obvious, effect of dualism is to protect science from its postmodern cultural adversaries, mostly in the humanities, who deny its claim to truth and objectivity. The dualism I have in mind says in effect to those who wish to export their radical skepticism to the sciences: "*noli me tangere*, do not encroach on the province of science about which you know very little, and science will not encroach on cultural studies." The mathematical physicist Alan Sokal submitted a nonsense-filled article that endorsed a radically skeptical view of scientific objectivity to *Social Text,* a journal of cultural studies with a postmodern bias. The hoax was exposed after publication. The alacrity with which the article was accepted for publication and the anger and defensiveness displayed by its editors betrayed their ideological animus and intellectual bankruptcy. Weinberg wrote in defense of Sokal for the *New York Review of Books* as a physicist and a dualist with a knowledge of his subject. The editors of *Social Text* were willing to appropriate to their monistic radical skepticism a field about which they knew little, if anything. Weinberg had the advantage of knowing what he was talking about.

As do Paul Gross and Norman Levitt in *Higher Superstition: The Academic Left and Its Quarrel with Science*, a devastating critique of postmodernist pretensions to knowledge about science. Gross is a distinguished professor of the life sciences, and Levitt is a mathematician. They demonstrate in case after case (the book reads like a trial of the intellectually ignorant and incompetent) how ignorant the critics engaged in postmodern "science studies" are of the sciences they wish to demystify. They point to elementary confusions, such as conflating topology and typography and adducing Heisenberg's Uncertainty Principle in the case against scientific objectivity. (Gross and Levitt point out that the Uncertainly Principle is "a predictive law about the behavior of concrete phenomena that can be tested and confirmed like other physical principles" [51-52].) The aim of the radical skepticism of "the academic left" (to be distinguished from the non-academic political left to which Sokal subscribes) is to erode confidence in scientific objectivity. It is fair to say that radical skepticism is an equal opportunity corrosive acid that applies to all the disciplines, scientific and non-scientific.

For the most part, Gross and Levitt are persuasive in their severity toward postmodern "science studies." What they fail to do, however, is to suggest any legitimate grounds for autocritique in the sciences. Since intellectual rigor and a respect for evidence are hallmarks of science, self-criticism should be part of its work. Are Gross and Levitt themselves betraying a scientistic hubris when they ask the following question, "If the few simple axioms adumbrated by the *Principia* could be induced to yield precise accounts of the orbits and comets, of the eccentricity of the earth and precession of its axis, of the pattern of oceanic tides, why should there not be an equally elegant, comprehensive, and reliable systematization of the study of human affairs?" (18). It is not at all clear whether the question is rhetorical, because, though they acknowledge that in hindsight scientifically motivated utopianism has produced failures that were "variously fatuous, quixotic or disastrous," they assert that "these disputes...are not central to our point." Why not?

Gross and Levitt are good writers, who come across as smart, knowledgeable, and self-assured. But there is a problem of tone and manner. From beginning to end, we are made to feel that not an inch will be ceded to their adversaries. All quotations from their work will be damning—as if there is nothing of value to be found elsewhere in their work. Unless the reader has matched or surpassed the authors in reading what they have read, he or she has no way of knowing with certainty how fair and judicious the authors have been. What the reader has to rely on is a sense of the unquestionable high intelligence and knowledge of the authors in scientific matters. I cannot presume to judge them on the scientific terrain. But when they speculate about the sources of the putative resentment of humanists of postmodern persuasion toward the sciences, they betray an unearned arrogance. They attribute the resentment to the effect of "the no-nonsense logical positivism adumbrated in such influential books as A. J. Ayer's *Language, Truth and Logic.*" The effect of logical positivism, they argue, was "devastatingly hurtful to the *amour proper* of traditional humanists," because in "impos[ing] severe tests of meaningfulness on all sorts of propositions," humanists were made to feel the meaninglessness of their own statements. "The propositions of science, by and large, escape humiliation, while those of the humanities, including such venerable philosophic areas as ethics and aesthetics, emphatically do not" (86-87). Growing up intellectually as I did in the humanities in the forties and the fifties of the previous century *before the advent of postmodernism,* I do not recall that I or my teachers, colleagues, and fellow students experienced humiliation at the hands of the logical positivists. To the extent that we were aware of positivism (and many humanists were) we either judged (to the extent that we were capable of judging) the philosophy as impoverished in its conception of truth, or at best relevant to the sciences, but not to the truths of the humanities. Shakespeare and Dante, Tolstoy and Dostoevsky suffered no humiliation in their reputations, so why should the scholars and critics who devoted their careers to understanding them? As cultural historians, Gross and Levitt leave something to be desired. Moreover, in characterizing logical positivism, they do not say where they stand on the question of its philosophical authority, though they are not loath to express forceful opinions elsewhere. They leave the reader hanging on the question of whether humanists *should have* experienced humiliation at the hands of the positivists. Given what precedes and follows in the book, this reader at least cannot resist the suspicion of *schadenfreude* at the expense of humanists, including those who are not postmodernists. The code of the dualist requires humility toward the views of those on the other side of the divide. Gross and Levitt are not dualists.

My own take on postmodern radical skepticism is that it mirrors the extreme and dogmatic scientism, which Wilson, Dawkins, Pinker, and others espouse. For Wilson, science and postmodern epistemology would appear to be the either/or of theoretical debate in the academy. Once again: "Postmodernism is the ultimate polar antithesis to the Enlightenment. The difference between the two extremes can be explained as follows: Enlightenment thinkers believe we can know everything, and radical postmodernists believe we can know nothing" (44). (Note the illicit shift from "postmodernism" to "radical postmodernism," in effect eliding a significant distinction. There are postmodernists who are not radical skeptics, believing as they do in local, if not universal, knowledge. What all postmodernists have in common is an aversion to all universalist claims.) We need not be radical skeptics to doubt that we can know everything. Wilson seems never to have tasted the fruit of skepticism, itself a product of the Enlightenment when it directs its fire against unwarranted confidence in certainties, its own as well as those of its adversaries. His confidence that everything can be brought into the light is unlimited.

My own admiring view of the Enlightenment is chastened by criticisms of its hubristic tendencies, but it is poles apart from the kind of critique one finds, for instance, in "post-colonial studies," an offshoot of postmodernism. I do not consider the Enlightenment or Western science culpable for the behavior of imperial powers when they suppress the ritual practice of *sati*, the self-immolation of widows in Hindu India or when they proscribe female circumcision in African countries. On the contrary, such behavior is in my view one of the benefits of the Enlightenment. Local knowledge and tribal rituals may have humanly destructive consequences. Modern scientific knowledge may be of Western origin but universal in its content and application when it contributes to the ridding of disease, the improvement of the environment, the prevention of and protection against natural disasters. The practices of the sciences in their areas of competence know no national, political, racial or gender boundaries. It is obscurantist and foolish to speak of an American or Russian science, a white or black science, a capitalist or socialist science, a male or female science. (We need only recall the chilling period in our history when the Nazis spoke contemptuously of modern physics as a Jewish science.) What is the case is that nation, race, economics, and gender may determine some of the tasks that science undertakes, but not its methods of experimentation or its logic. Scientism, in my pejorative use of the term, refers to its tendency to move beyond science's area of competence. In the process it ceases to be a science while pretending to be one.

In Wilson's version of intellectual history, Enlightenment theory aims for the unity of all knowledge, the finding of ultimate explanations for everything, the opposing theory declares for a radical skepticism about the possibility of any knowledge. Missing from the quarrel is the excluded middle. Both theories then are grand theories (radical postmodernists would bridle at the attribution) with the ambition to account for everything. Both theories are dogmatic and therefore incapable of that mixture of confidence and epistemological modesty which says, "this we know, this we can know, this remains in the realm of mystery subject to a variety of speculation and interpretation that cannot and perhaps never will be resolved to certain knowledge." Perhaps it is the case, as Berlin has remarked, that "it is…easier to exaggerate, to lean to an extreme" (265). Exaggeration improves the chances that what one says will be remembered.

It is astonishing that despite the evidence of the amazing progress sciences have made in, for example, "our" understanding of the genetic makeup of living creatures, radical skeptics refuse to grant their discoveries the status of objective knowledge. (I place "our" in quotation marks because we delegate scientific understanding and conviction to scientists. Our faith or trust in them is based in part on the evidence of the technology and medical advances that have come out of science.) It is of course true that scientific claims may be provisional and can be superseded by new knowledge, but there are claims that have been consolidated and have not been superseded and even those claims that have been superseded can be placed on a curve of progress to a better understanding of phenomena. A belief in the possibility of objective knowledge is necessary to scientific practice.

*   *   *

If Weinberg's dualism acknowledges the limits of science with respect to other disciplines, it is less modest about what it can accomplish in understanding the nature of the physical universe. He cites the contrary view of Karl Popper, "the dean of modern philosophers of science," who rejects the very "idea of an ultimate explanation." Between Weinberg and Popper on the question of a final theory, I would side with Popper. According to Popper, "every explanation may be further explained, by a theory or conjecture of a higher degree of universality. There can be no explanation that is not in need of a further explanation" (*FU*, 230). In a recent review in the Times Literary Supplement of Susan Haack's *Defending Science: Within Reason*, a book which is critical of Popper, the sympathetic reviewer, Roger Kimball, makes it seem as if Popper is a postmodern radical skeptic about scientific truth. "What was novel about Popper's doctrine was not the idea that negative instances disconfirm or 'falsify' a theory but rather the amazing thought that positive instances do not—in principle *cannot*—act to confirm a proposition or theory. Scientific laws, he says, 'can never be supported, or corroborated, or confirmed by empirical evidence'" (9). Amazing it is not. Haack is careful to make a distinction between corroboration and confirmation. An experiment can corroborate a hypothesis or a theory without confirming it. Though she is critical of what she views as inconsistencies in Popper's position, she is respectful of his work and is certainly not amazed (in the negative sense) by it. Popper is in the line of Humian skepticism that supports probability rather than certitude, not as a weapon against science, but as an argument for the openness of scientific inquiry? No experiment can absolutely prove a theory since the future can never be demonstrated with absolute certainty; all it can do is show it to be true to a high degree of probability. If the validity of scientific claims depends upon induction, no generalization based on evidence acquired through induction could possibly exhaust all the instances past, present and future that would *absolutely* confirm or disconfirm the generalization. Haack notes that Popper is not alone in his views. She points to his affinities with (as well as differences from) other philosophers of science such as Thomas Kuhn and Imre Lakatos. It is clear from her exposition that Popper is a friend of science and not a radical skeptic about its claims. According to John Stuart Mill, "The beliefs we have most warrant for, have no safeguard to rest on, but a standing invitation to the whole world to prove them unfounded" (quoted in Aileen Kelley 114). Mill speaks of beliefs, but this could apply to scientific claims as well. The implication here is not radical skepticism, but rather a doubt about whether one has arrived at truth at any moment in time while at the same time affirming its existence. Mill here anticipates Popper's theory of falsification.

In his essay, "The Corroboration' of Theories," the philosopher Hilary Putnam also quarrels with Popper's view that scientific laws are *falsifiable*, not verifiable. According to Putnam, he is less concerned with the truth of a theory than with its vulnerability. His attitude is in principle skeptical. He is interested in testing theories, not in their practical application. "Failure to see the primacy of practice...leads Popper to the idea of a sharp demarcation between science, on the one hand, and political, philosophical, and ethical ideas on the other. This 'demarcation' is pernicious, in my view; fundamentally, it corresponds to Popper's separation of theory from practice, and his related separation of the critical tendency in science from the explanatory tendency in science" (Hacking, ed. *Scientific Revolutions*, 78). Putnam takes a dim view of the consequences of this separation. "The failure to see the primacy of practice leads Popper to some reactionary conclusions. Marxists believe that there are laws of society; that these laws can be known and that men can and should act on this knowledge. It is not my intention to argue that this Marxist view is correct; but surely any view that rules this out *a priori* is reactionary. Yet this is precisely what Popper does—and in the name of an anti priori philosophy of knowledge" (79). So he takes Popper to task on two grounds: his philosophical inconsistency and the putative reactionary character of his philosophy.

Putnam provides no substance to the charge that Popper's view is reactionary. It comes across as name calling, unworthy of philosophical discourse. Moreover, there is something disingenuous in his unwillingness to examine the Marxist view while at the same time using it as a club to beat Popper. One might say in rejoinder that there is evidence that "the laws of society," however and by whomever they may be formulated, when applied, tend to create despotic and totalitarian societies. Putnam simply misreads the relationship between theory and practice in Popper's work. The demarcation between science and political, philosophical and ethical ideas that he postulates is based on practical observation of the pernicious effects of applying "laws of society," usually (I am tempted to say, always) inadequate to the complexities of human life. Popper's demarcation is meant to serve the *practical* interests of "an open society." The charge of philosophical inconsistency seems to me trivial, given what is at stake. We can excuse "inconsistency" when it serves the greater consistency of Popper's devotion to openness. "The laws of nature" are one thing, "the laws of society" another. The danger of the latter is the closure of tyranny.

Weinberg has no interest in formulating "laws of society," but he dreams of a final theory of the physical universe. Popper rejects its possibility for reasons similar to those of his rejection of a scientific theory of society: every theory is vulnerable to refutation since new evidence may always come into view and the declaration that a final theory has been achieved may have a chilling effect on future inquiry and discovery.

Weinberg does not have as yet a final theory. So what is the status of the scientific knowledge that he possesses? A word that recurs like a refrain in his essays is "approximation." Maxwell's equations for electricity and magnetism are "approximations that are valid in a limited context" (*FU*, 50). Weinberg does not "regard approximations as mere useful fictions... [A]pproximate theories are not merely approximately true. They can make a statement that, though it refers to an approximation, is nevertheless precisely true" (208). (I don't know why "precisely" is needed in the sentence.) Apropos of the objective truth of quantum mechanics, Weinberg has this to say: "quantum mechanics may survive not merely as an approximation to a deeper truth, in the way that Newton's theory of gravitation survives as an approximation to Einstein's general theory of relativity, but as a precisely valid feature of the final theory" (*Dreams*, 88-89). Knowledge, on this understanding, is probable and provisional, not final or absolute. Certain kinds of knowledge have such a high degree of probable truth that one can allow the scientist something close to certainty about it. For Weinberg, objective truth lies in the progress toward the final truth. In Aristotelian terms, scientific activity is best understood not in terms of efficient causes (historical conditions and circumstances, personal bias, fashion, the view of Thomas Kuhn), but rather in terms of the final cause. Scientific progress, for Weinberg and the likeminded, has its telos in the final truth. Weinberg acknowledges that Kuhn "does not deny that there is progress in science," only that for Kuhn there is no set goal to the progress. "It is a progress driven from behind, rather than pulled to some fixed goal to which it grows ever closer" (*FU*, 19). Absent from Weinberg's account is the subjective element, which we find in Niels Bohr's understanding of scientific inquiry. As Gerald Holton reports, to the question of whether the world can "be known more and more certainly, independent of our own predilections, or decisions, our laboratory arrangements," Bohr's answer would be a "resounding 'No.'" "Objective knowledge of a phenomenon, in Bohr's terms, is what you learn from the full report of the experimental arrangements that probe into the phenomenon—arrangements, be it noted, of apparatus on the scale of everyday life, and describable in ordinary language (with mathematics merely a compact and refined extension of it). There is no firm boundary between that which is observed and the observing machinery" (*Thematic Origins*, 465). By introducing a "subjective" element, Bohr means to enrich, not undermine, the objectivity of science. Subjectivity adds to, does not subtract from, our knowledge of the world.

The value of a theory does not depend exclusively on its approximate correspondence to reality. Perhaps the most striking and most unexpected aspect of Weinberg's argument for scientific objectivity is his insistence on the aesthetic aspect of scientific theory. He is not, of course, the first or the only scientist to insist on the beauty of scientific discovery, but many see it merely as a side effect, a pleasure incidental to the truth of the theory. Weinberg, in contrast, views the aesthetic aspect as essential to the truth value of the theory. And he is careful to specify exactly what he means. The aesthetic of science is not to be confused or conflated with the aesthetic of art or literature. Nor is it elegance, as in Brian Greene's "elegant universe." Weinberg refers to Einstein's comment about leaving elegance to tailors. The beauty of theory is in the simplicity of its ideas, the "sense of inevitability that the theory gives us," its obedience to "principles of symmetry...not the symmetries of things, but the symmetries of laws," and, perhaps most surprisingly, "rigidity" (*Dreams*, 134-37, 147). Weinberg speculates that "general acceptance of the general theory of relativity was due in large part to the attractions of the theory itself—in short, to its beauty" (98). In any event, without a final theory Weinberg is left at least for the present with *approximation*.

Modern physics, lacking a final theory, has operated productively under the aegis of two incompatible theories: relativity theory (for the macrocosm) and quantum mechanics (for the microcosm). Scientists may seek unity, but not having achieved it, they entertain plural theories. Relativity theory and quantum mechanics are both pragmatically fruitful and yet theoretically incompatible. Could it be that the potential incompatibility of theories (both possessing approximations of the truth) is a way of keeping scientific inquiry always open? Bohr, a founder of quantum theory, sees the ambition to achieve a final theory as misguided. He addresses the problem of antithetic theories, both of which are valid, by introducing the principle of complementarity. In *Thematic Origins of Scientific Thought*, Gerald Holton formulates the problem and Bohr's solution in the following manner:

> The puzzle raised by the gulf between the classical description and quantum description was: could one hope that, as had happened so often in physics, one of the two antithetical views would somehow be subsumed under or dissolved in the other (somewhat as Galileo and Newton had shown celestial physics to be no different from terrestrial physics)? Or would one have to settle for two so radically different modes of description of physical phenomena? Would the essential continuity that underlies classical description, where coordinates such as space, time, energy and momentum can in principle be considered infinitely divisible, remain unyieldingly antithetical to the essential discontinuity and discreteness of atomic processes? (101)

Bohr's proposal was essentially that we should attempt not to force a reconciliation of antithetical theories, but rather to realize the complementarity of representations of events in these two quite different "languages" (102). *One might say that Bohr qualifies as kind of dualist within the community of physical scientists.*

Complementarity, a substitute for a final theory, means that the two languages are not "to be subsumed or dissolved in the other." The inspiration for complementarity, Holton tells us, came not from science, but from Bohr's teacher, the philosopher Harald Hoffding, an admirer of both William James and Søren Kierkegaard. Bohr found complementarity in James, who concluded from his psychological research that "in *certain persons*, at least, *the total possible consciousness may be split into parts which coexist but mutually ignore each other,* and share the objects of knowledge between them. More remarkable still, they are *complementary*. Give an object to one of the consciousnesses, and by that fact you remove it from the other or others. Barring a certain common fund of information, like the command of language, etc, what the upper self knows the under self is ignorant of, and *visa versa*" (*Thematic Origins*, 125-26). And then there is the surprising influence of Søren Kierkegaard, whose leading idea, according to Hoffding, "was that different possible conceptions of life are so sharply opposed to one another that we must make a choice between them" (130). Against the romantic conception of the continuity of life, Kierkegaard espoused the idea of discontinuity in which consciousness leaps back and forth between possibilities.

Complementarity is a philosophical theory, not a scientific theory like relativity or quantum mechanics. It is the result of the influences of two philosophers, one of whom (Kierkegaard) was not especially sympathetic to the scientific enterprise.

Its advent in Bohr's thought should be a caution against the views of Wilson and Dawkins, who find little instructive in philosophy, envisaging as they do its disappearance into the sciences. Consider these remarks by Einstein: "The reciprocal relationship between science and epistemology is of a noteworthy kind. They are dependent on each other. Epistemology without contact with science becomes an empty scheme. Science without epistemology is—insofar as thinkable at all primitive and muddled" (quoted in *Thematics*, 355). Consilience, if it is to be achieved, is not a one way affair. It is not systematic, occurring as it does from time to time. Without a multiplicity of disciplines, each following its own imperatives, though always open to fruitful influences from other disciplines, the closed-minded imperialism of a discipline, of any discipline, is ultimately doomed to sterility.

Weinberg does not address the challenge of complementarity to his dream of a final theory. He concedes that Popper and others "may be right," though he refuses to give up the possibility of a final theory on logical grounds. "But I do not think that this position can be argued on the grounds that no one has yet found a final theory. That would be like a nineteenth-century explorer arguing that, because all previous arctic explorations over hundreds of years had always found that however far north they penetrated there was still more sea and ice left to the north, either there was no North Pole or in any case no one would ever reach it. Some searches do come to an end" (*Dreams*, 230-31). I am in sympathy with Weinberg's dualism as it relates to the relations between the natural sciences and human affairs. As for the dream of a final theory, my inclination is to side with Popper and Bohr against Weinberg, because it closes off the possibility that new fundamental discoveries may be made or that new understandings can be achieved. The idea of a final theory may undo the healthy skepticism that accompanies a belief in the probabilistic nature of truth. At the same time, it would be a mistake to discourage those who pursue a final theory not to persist in their pursuit.

The quarrel may turn into a dialogue in which the two sides cast the differences in terms that allow a place for both views, for complementarity, so to speak. In a review of a book by Brian Greene in the *New York Review of Books*, Freeman Dyson divides physicists info the revolutionary camp and the conservative camp. "In the history of science there is always a tension between revolutionaries and conservatives, between those who build castles in the air and those who prefer to lay one brick at a time on solid ground" (16). The revolutionaries (Einstein et al.) dream of a final theory, the conservatives (Rutherford et al.) are willing to accept the complementarity of separate theories. What may be surprising, one might say counterintuitive, about this division is that it is the revolutionaries who seek to close off the future of their discipline in a final theory, while the conservatives want to keep the future open. Like Weinberg, Greene with his string theory is on the revolutionary side, Dyson on the conservative. Referring to a debate that had taken place between them, Dyson says, "I ended by saying that I rejoiced in the fact that science is inexhaustible and I hoped the nonscientists in the audience would rejoice too" (19). And yet it would be folly to have this debate end with an unequivocal victory for either side. Dyson notes that "there has always been a tension," and scientists have worked off that tension. Even if the final theory proves to be elusive, the work done in trying to formulate it will doubtless yield new insights into the physical world. A complete victory for the conservatives would be at the unfortunate expense of the greatest scientists, Einstein, Schrodinger, Heisenberg, among others. The grand system builders have the most honored place in the scientific community. Nevertheless, conservative resistance to the quest for final closure can be salutary as a caution against hubris and a guarantee that scientific inquiry will continue. (There are the grand system builders in the social sciences and in the humanities, but they are fortunately far and few between, especially in the humanities.) Certainly most humanists would rejoice with Dyson in his view of the openness and inexhaustibility of knowledge.

The desire for a final theory may be inherent in the scientific enterprise, which seeks to understand as much as possible. Indeed, progress in science may require the goad of belief that total understanding is possible—even if it may not be. It may be a paradox of the health of science that the belief in the possibility of total understanding should be coupled with doubt that it can ever be actualized. Who can ever be sure that the conviction that one has finally formulated the laws of nature is not premature? Or differently put, such a conviction may prove an obstacle to testing and having to go beyond these laws. Where "the laws of society" are concerned, the question arises: how should one apply them? It takes little imagination and empirical observation to see the risks of tyranny, given the uncertainty about their reliability in the application.

* * *

Science has its external adversaries: creationists who turn to scripture or church dogma and refuse to take seriously the evidence of evolutionary theory, mystics for whom matter is an illusion and therefore not worthy of study, romantic poets who oppose imagination to reason. The most insidious of enemies (because least visible) may be the enemy within who appears as a friend. Scientism is such an adversary. In the name of science, it promises what is inimical to the scientific enterprise, the conviction of certainty. There is no practicing scientist or theorist of science more sensitive to the dangers of certainty than Richard Feynman. Again and again, he affirms the attitude of doubt, of persistent doubt, as essential to scientific inquiry. Doubt, I should add, not as the radical skeptics would have it to undermine confidence in scientific discovery, but rather in its service. Science "teaches the value of rational thought, as well as the importance of freedom of thought; the positive results that come from doubting that the lessons are all true" (186). Addressing the subject of the teaching of science in *The Pleasure of Finding Things Out*, Feynman remarks that "science doesn't teach [that something is 'such and such'] experience teaches it" (187). And he goes on to speak of our "unscientific age" in which books purporting to be scientific are in fact unscientific. "As a result, there is a considerable amount of intellectual tyranny in the name of science" (188). If religions subsist on the certainty that God exists, the scientist in his capacity as scientist can have no part in them. (Feynman knows, of course, that there is more to religion than the belief in the existence of God and he respects that more, but his quarry here is the conviction of certainty.) "The uncertainty that is necessary in order to appreciate nature is not easily correlated with the feeling of certainty in faith, which is usually associated with deep religious belief. I do not believe that the scientist can have that same certainty of faith that very religious people have." And then in the spirit of uncertainty he makes the qualification: "Perhaps they can. I don't know" (*The Meaning of it All*, 43). Scientism with its certainties turns science into a religion and thus betrays its essential character. Science, according to Feynman, not only requires the attitude of doubt, it must also have a sense of its limits. Thus, though "science can be defined as a method for, and a body of information obtained by, trying to answer questions put into the form: If I do this, what will happen?" the judgment of the ethical choices one makes and the actions one takes is outside the purview of science (*The Pleasure*, 255). "I don't believe that a real conflict with science will arise in the ethical aspect, because I believe that moral questions are outside of the scientific realm" (254). I should think that at the time that Feynman wrote this the statement would have already been a truism, but his need to make the statement suggests that among his colleagues it might still be a controversial thing to say. My decision to quote it suggests that ethics is still not safe from scientism.

* * *

We have no reason to despair if the disciplines and the arts have each its own imperatives, which may or may not converge. The prospect of disciplines going off in different directions, unconstrained by the demand for consilience, may bring greater rewards in the intellectual field than the illusory prospect of the unity of knowledge. Indeed, a serious reflection about the differences between the scientific and literary enterprises, for example, reveals how fundamental they are and how necessary for the sake of our knowledge of the world to preserve those differences. From the perspective of the "hard sciences" (physics, chemistry, even biology), the disciplines of the humanities are "soft." Should literary scholars and historians feel troubled about it and try to harden their disciplines in the spirit of the natural sciences? I think not. A discipline may be soft not because it is immature, but because of the complexity and resistance of its subject matter. The hardening of the discipline is too often accompanied by disrespect for the subject.

There are also attitudinal differences between disciplines that need to be understood and respected. In "The Two Cultures," C. P. Snow makes a distinction between science as forward looking and the literary imagination as backward looking. If we set aside its unnecessary invidiousness, we can fruitfully explore the distinction. Practicing scientists are for the most part not concerned with the history of science—at least not in their professional capacity. They may have curiosity about it, but that curiosity need not come into play in their work as scientists. The history of science may have no intrinsic interest for their work. Each theory surpasses or is intended to surpass predecessor theories in their explanatory power. A theorist should know the work of his immediate predecessors, but that is different from needing to know Aristotelian or Ptolemaic science. In contrast, neither literary achievement nor the scholarly and critical understanding of it is cumulative in the progressive sense. Literary works that survive have at once an existence in a non-progressive continuum of tradition and an enduring life of their own. (Eliot does not improve upon Donne, nor does Shaw upon Shakespeare.) Within the tradition there is influence; which is in effect a tribute of the new work to its predecessor. Harold Bloom's controversial theory of "the anxiety of influence" asserts the existence of agonistic relations between the latecomer poet and his predecessor, but even there the antagonism itself implies immense respect for the power of the work of the predecessor. Influence in the literary tradition may not come from the immediate literary past. T. S. Eliot skipped over his Victorian, Romantic, and Augustan predecessors to the metaphysical poets of the seventeenth century to find inspiration and validation for his own creative and critical work. Yet he possessed the immediate literary past, indeed the whole of the literary tradition. A literary critic may not share the religious or philosophical assumptions of a work of the past (the *Iliad, The Divine Comedy* and *Paradise Lost* come to mind as subjects for the modern skeptical, agnostic poet or reader), but having been taught to "suspend disbelief" (Coleridge's phrase), he or she can enter sympathetically into the imaginative vision of the work and discover or recover its power.

Thomas Kuhn's *The Structure of Scientific Revolutions*, without intending to, casts light on the difference I am setting out between the scientific imagination and the literary imagination. John L. Casti provides a vivid account of how Kuhn came to his "notion of a scientific *paradigm*." In preparing for a course on the origins of seventeenth-century mechanics, Kuhn went back to Aristotle's physics, which held that "all matter was composed of spirit, form, and qualities, the qualities being air, earth, fire and water."

> Kuhn wondered how such a brilliant and deep thinker, a man who had single-handedly invented the deductive method, could have been so flatly wrong about so many things involving the nature of the physical world. Then, as Kuhn recounts it, one hot summer day the answer came to him in a flash while he was poring over ancient texts in the library: Look at the universe through Aristotle's eyes! Instead of trying to squeeze Aristotle's view of things in a modern framework of atoms, molecules, quantum levels, and so forth, put yourself in Aristotle's position, give yourself the prevailing worldview of Aristotle's time, and lo and behold, all will be light. For instance, if you adopt Aristotle's worldview, one of the presuppositions is that every body seeks the location where by its nature it belongs. With this presumption, what could be more natural than to think of material bodies as having spirits, so that "heavenly" bodies of airlike quality rise, while the spirit of "earthly" bodies causes them to fall? (39).

In a sense, Kuhn was here performing a role comparable to that of the Coleridgean literary critic. He was suspending disbelief in order to enter and experience Aristotle's mental universe. Could this be part of the explanation of why Kuhn's work has provoked the ire of scientists and the sympathy of humanists? In a quite different spirit, when Richard Dawkins grudgingly concedes the historical interest of thought about the questions of the purpose and meaning of life in work done before

1859 while declaring the objective worthlessness of its answers to those questions, he is making a concession to a line of inquiry that he has no real interest in pursuing. Unlike Kuhn, he seems to have little, if any, sympathy for the exercise of the historical or literary imagination.

The boundaries between disciplines are not and should not be impermeable. Science and scientists have been and will continue to be subjects of the literary imagination. Lydgate in George Eliot's *Middlemarch* has the heroic ambition of trying to discover the primal tissue that is the basis of life. Thomas Mann's *The Magic Mountain* would be impoverished without its preoccupation with medical practices in the sanatorium at Davos. And Ian McEwan in his recent novel *Saturday* describes the skills of its neurosurgeon protagonist with an almost reverential meticulousness. Indeed, the novel is in part about the fruitful rivalry of the scientific sensibility of the protagonist and the literary sensibility of his poet daughter. The relationship between the science and the humanities may be adversarial or dialogical or dialectical (they should be able to learn from each other), but each exists in its own sphere; disciplines (and their exponents) may respect one another's autonomy while learning from them. I am not entirely comfortable in calling myself a dualist. I would prefer a term that would cover the overlappings and continuities between disciplines that would at the same time admit fundamental differences between them not to be absorbed into a factitious unity of knowledge.

Epilogue

I have taken on scientism with the apprehensiveness that I will be providing ammunition to the anti-scientific fundamentalists who say that Darwinism is not a fact but a theory—as if theories and facts are opposites. A theory explains facts; it does not oppose them. I would like to think that my argument against Darwinian misadventures in the humanities is not only in the interest of the humanities, but also in the interest of science itself, for hubris and self-misrepresentation can only bring harm to its agents. As I have made clear, I do not ally myself with the practitioners of postmodern "science studies" because their dogmatic skepticism does a disservice to science.

I would also separate myself from scientism's severest critic, the philosopher of science Paul Feyerabend, who at times makes no distinction, as I do, between science and scientism. One of his essays is titled "How to Defend Society Against Science," in which he declares: "I want to defend society and its inhabitants from all ideologues, science included" (Hacking ed. *Scientific Revolutions*, 156). According to Feyerabend, methods and results that are supposed to separate the sciences from the non-sciences are myths, no different from the myths from which science is supposed have liberated us. He wants to erode the boundaries between science, on the one hand, and religion and poetry, on the other, to the advantage, I believe, of the latter. For him, the privileging of science and the scientist in modern society is elitist, and elitism is undemocratic and hence anathema. The bane of the "progress" of the sciences is specialization and the cult of the expert, which removes the layman from the court that decides the value of scientific work. Feyerabend speaks approvingly of "democratically elected consulting bodies" (162).

It is difficult to believe that he would want the laity to decide the truth of a scientific theory or discovery. Presumably he would want it to judge its value for society. He may be right that the expert, unchecked by others, represents a threat to democracy, and indeed, if he or she is corrupt, to the truth. He does not entertain the possibility that experts checking on other experts, as if there is an inherent corruption in expertise. Nor does he seem concerned with the question of the competence and integrity of the laity. His "consulting bodies" bring to mind juries, who do not require an expert knowledge of the law to decide questions of guilt and innocence. (Would he be concerned about jury nullification?) Consistent with his animus against the expert is his aversion to scientific jargon, which he views as part of the effort of the expert to maintain control of his or her discipline. Like Orwell before him, Feyerabend has caught hold of an obscurantist disciplinary tendency (of course, not confined to the scientific disciplines), but he writes at times as if the tendency constitutes the essence of the discipline. He also takes to task scientists for their arrogant assurance that their methods and results make it possible for them to understand and master the world. His prose is marked by indignant passion and the hyperbole that springs from passion. He tends to assert or declaim rather than to argue and demonstrate. He is right to worry about the kind of specialization that confines the mind to a particular body of knowledge and to a particular perspective, but does he believe that scientific progress is possible without specialization?

I would not wish to serve on the kind of consultative body that Feyerabend envisages as a check on the work of scientists. I leave science to the scientists except where it extends itself into areas that it cannot comprehend, or where it is indifferent to what is already known. I think of sociobiology and evolutionary psychology roughly on the analogy of nineteenth-century liberal economic theory. It was a powerful theory, but it came to be called the dismal science because of its (willful?) blindness to the immiseration of the masses that its doctrine seemed to justify. Sociobiology and evolutionary psychology for all their genuine accomplishments risk incurring the same judgment because of their impoverished view of human culture. The theory of evolution is a great achievement, and the assault against it by creationists a national scandal. But the cause of science is not served when prominent practitioners overreach in their claims for the theory.

Since I am defending the humanities, I should say something about their exponents. "Humanists" may refer to those like myself who practice the disciplines under the rubric of the humanities, or the term may refer to those who have a philosophy or way of thinking called humanism. The provenance of humanism is the Renaissance, its exemplary figure Erasmus. In the time of religious conflict between the established Church in Rome and the Reformation, humanism represented a return to the human-centered world of the Greek and Roman classics. In the humanist perspective, "man is the measure," in Erich Kahler's phrase. Erasmus may not have denied Christian revelation, but his heart and mind were elsewhere. In its human-centeredness, humanism has an affinity for science, and one might even count the sciences among the humanities.

Science and the humanities were not always at "war," a hyperbolic metaphor. From ancient times to the Renaissance the distinct fields of knowledge first laid out by Aristotle existed on a continuum essentially undisturbed by the encroachments of one field upon another. In the Middle Ages, the continuum became a hierarchy with theology assuming the role of queen of the sciences. The seeds of usurpation were present in scholasticism; reason exercised in behalf of faith threatened to assert its autonomy. The long peace was undermined by the Copernican revolution, which replaced Ptolemaic earth-centered cosmology with a heliocentric one and by the confirming experiments of Kepler and Galileo. But all three scientists continued to do obeisance to the authority of the Church. It is Descartes who finally frees science from that authority. Cartesian dualism breaks apart the spiritual and material worlds. The material world, released from its bonds to theology, becomes subject to the rule of science. According to Cartesian scripture, we must render to God that which is spirit and to science that which is matter. The supreme genius of the Enlightenment, of course, is Newton. A devout man, he was uncomfortable with dualism as was the philosopher Spinoza. Newton reunited the material world with God, turning him into the maker and regulator of the machine that runs the cosmos. But by bringing the Creator down to the creation, he effectively contributed to the decline of religious authority in post-Enlightenment Europe, leaving a vacuum for those whose spiritual needs found no satisfaction in a mechanical universe. Darwin's revolutionary account of the origins of the human species directly contradicted the Biblical account of origins and *for many* delivered the coup de grace to the literalist version of biblical authority.

For many, but not for everybody. One effect of Darwinism in the nineteenth century, which has persisted through the twentieth and the present century, has been to intensify the defensive literalism of the fundamentalist interpretation. It is almost as if the church enters the fray on the scientist's ground of factual truth and finds itself at a disadvantage. The corrosive rationalism of a writer like Thomas Paine in *The Age of Reason* exposes the inconsistencies and absurdities of literalist credulity. A scientifically inspired higher criticism, more sensitive to the symbolic truths in the Bible than Paine's ruthlessly rationalistic criticism, is even more effective in undermining fundamentalist interpretation. Matthew

Arnold's formulation in "The Study of Poetry" (1880) is memorable. It can serve as a motto for nineteenth-century humanism. "There is not a creed which is not shaken, not a received tradition which does not threaten to dissolve. Our religion has materialized in the fact, the supposed fact [i.e. the belief in the literal truth of the Bible]; it has attached its emotion to the fact, and now the fact is failing it" (161). As an answer to the question of what would be taking its place in the wake of the decline of religion, Arnold proposed poetry, the truth of which does not lie in its literalness. With poetry in authority, the Bible itself could be redeemed for the imagination. "The future of poetry is immense, because in poetry, where it is worthy of its high destinies, our race, as time goes on, will find an ever surer and surer stay.... Poetry attaches its emotion to the idea, the idea is the fact. The strongest part of our religion today is its unconscious poetry."

Arnold is in the line of Blake and Wordsworth, who affirm the spiritual authority of poetry. The rival contender for authority was the science of Newton and Locke, which Blake disparagingly characterized as "natural religion." For Blake, modern science was not simply a description of things as they are, it was a reduction of the world to its material aspect and therefore a debasement of it. The senses alone provide an inadequate access to the truths of the world, which only the imagination can realize. Rather than repudiating the senses, Blake wants them integrated into the life of the imagination. Less combative than Blake, Wordsworth presents a more benign view of the relations between poetry and science in his Preface to *Lyrical Ballads*:

> The knowledge both of the poet and the man of science is pleasure; but the knowledge of the one cleaves to us as a necessary part of our existence, a natural and unalienable inheritance; the other is a personal and individual acquisition, slow to come to us, and by no habitual and direct sympathy connecting us with our fellow-beings. The man of science seeks truth as a remote and unknown benefactor; he cherishes and loves it in his solitude: the poet, singing a song in which all human beings join with him, rejoices in the presence of truth as our visible friend and hourly companion. Poetry is the breath and finer spirit of all knowledge; it is the impassioned expression which is the countenance of all science. (80-81)

Wordsworth avoids the language of conflict. Both science and poetry have a stake in knowledge and pleasure. The potential for conflict, however, is implicit in the invidious claim that he makes for poetry. Science deals in invisible structures in a language accessible only to those who do the work of science, whereas poetry deals in sensuous impression and feeling in a vernacular available to everyone. Poetry is the medium of fellow feeling in a way that science can never be, a claim that need not imply that science is an inferior activity: it could simply mean that science and poetry have different tasks, a view that scientists might find congenial. But Wordsworth goes further in speaking of poetry as "the breath and finer spirit of all knowledge," as if science and poetry are on a continuum in which poetry has the privileged place. Is this a claim made in a mood of rhetorical enthusiasm that would be hard to sustain in rational argument, or is there a case to be made for the poet's competitive access to the body of knowledge that is the province of science? Wordsworth's manifesto tells us nothing of the nature of either discipline, only of their putative social effect or affect. Though the language is one of conciliation, there is implicit in it a resistance to the increasing cultural ascendancy of science in the modern world. Shelley, a second-generation romantic poet, who was friendly to science, nevertheless designated the poet as the "unacknowledged legislator of the world." The conflict between science and the humanities has its origins in romanticism.

Perhaps no episode in modern cultural history better illustrates the gulf that is emerging in the romantic period between science and poetry than the scientific researches of the great German poet and man of letters, Johann Wolfgang von Goethe, and in particular his controversial anti-Newtonian theory of color. The distinguished German scientist Hermann von Helmholtz summed up the non-negotiable difference between Goethe's "method" and Newton's in a way that goes to the heart of the difference between poetic imagination and scientific investigation:

> In writing a poem, [the poet] has been accustomed to look, as it were, right into the subject, and to introduce his intuition without formulating any of the steps that led him to it. And his success is proportionate to the vividness of the intuition. Such is the fashion in which he would have nature attacked. But the natural philosopher [i.e. scientist] insists on transporting him into a world of invisible atoms and movements, of attractive and repulsive forces, whose intricate actions and reactions, though governed by strict laws, can scarcely be taken in at a glance. To him the impressions of sense are not an irrefragable authority; he examines what claim they have to be trusted; he asks whether things which they pronounce alike are really alike, and whether things that they pronounce different are really different, and often finds that he must answer, no! (*Science and Culture*, 13)

Goethe's resistance to Newton reflects a mistrust of scientific abstraction in which what is real and lawful in nature is inaccessible to the senses. It is the obverse of scientism in that it implicitly wants the poetic imagination to aggrandize all of reality—a kind of rear guard action against the advance of modern science. It is useful to keep in mind the aggrandizing ambition of romanticism (for that is what Goethe exemplifies in his scientific researches, not all of which were failures) as we consider the aggrandizing ambition of modern science.

We should not conclude from the contrast between Goethe and Newton the view that scientists and poets as human beings are a different species. Peter Medawar in his *Advice to a Young Scientist* presents a persuasive view of the variety of temperament to be found among scientists:

> Scientists are people of very dissimilar temperaments doing different things in very different ways. Among scientists are collectors, classifiers and compulsive tidiers-up; many are detectives by temperament and many are explorers; some are artists and others artisans. There are poet-scientists and philosopher-scientists and even a few mystics. What sort of mind or temperament can all these people be supposed to have in common? *Obligative* scientists must be very rare, and most people who are in fact scientists could easily have been something else instead. (3)

A scientist may indeed be also a poet, but doing science and writing poems are qualitatively different activities. In performing one or the other role, he or she is not a hyphenate. What this may mean is that the mind or brain is modular and that unusual breed the poet-scientist has the advantage of operating alternatively in either mode. Goethe is a special instance of a poet attempting to be a scientist in the poetic mode.

In *Science and the Modern World* (1925), Alfred North Whitehead attributed the romantic reaction against modern science to its division of objective reality into primary and secondary qualities. Primary qualities are mass and density, the invisible subject matter of scientific investigation; secondary qualities are color, sound, touch: the sensory subject matter of poetic imagination. Whitehead was on the side of the poets. Nature is not, as Locke in the spirit of Newtonian physics believed and as Whitehead phrases it, "a dull affair, soundless, scentless, colourless" (79). On this view, science inherits the Socratic (philosophical) mistrust of poetry and the belief that truth is to be found in the invisible structures that underlie the phenomenal world. Whereas poetry delights in phenomenal life, modern science takes no interest in it, indeed sees it as the source of illusion and falsehood. For Whitehead the resolution of the conflict could come only with the restoration of the integrity of the object world, though he did not say how this would be accomplished, how, that is, the kind of division between science and poetry that Helmholtz spells out can be overcome.

If the laws of science are invisible, its social effect becomes manifest and visible during the Industrial Revolution. Victorian writers (Carlyle, Ruskin, Dickens, and Morris), already anticipated by Blake's vision of the dark Satanic mills, severely criticize the degradation of man and landscape caused by the machine, a product of science. The industrial city is a soulless world of grinding machines and Gradgrinding fact. With the decline of religious authority in the nineteenth century, the prophetic voice is the poetic voice asserting itself against the demonic excesses of machinery, an emanation of scientific intelligence. Walter Pater, late in the nineteenth century, speaks hyperbolically of "all great poetry" as "a continual protest" against "the predominance of machinery" (61). Note that the protest is not against the machine per se, but against its "predominance." The tone of the criticism varies in intensity, depending on the temperament of the critic. Carlyle's denunciations of mechanism, for instance in "Signs of the Times" (1829), often have the sound of a Luddite jeremiad, but underneath his vatic fervor one hears another voice, that of conciliation. "Meanwhile, it seems clear enough that only in the right coordination of the two [the mechanical and the dynamical or the spiritual], and the vigorous forwarding of *both*, does our true line of action lie. Undue cultivation of the outward, again, though less immediately prejudicial, and even for the time productive of many palpable benefits, must, in the long-run, by destroying Moral Force, which is the parent of all other Force, prove not less certainly, and perhaps still more hopelessly, pernicious" (73). The post romantic claim is focused more on the machine, an application of science, than on science itself.

The romantic ambition for poetry has not been realized. In our time, poetry is a private religion for the few. Even its great modern exponent, T. S. Eliot, chastised those who presumed to conflate poetry with religion. James Joyce knew by the time he wrote *Ulysses* that Stephen Dedalus's high-flying aspiration to recreate the conscience of his race was aesthetic hubris. To change the world, one needs to create institutions, and poetry, which is personal expression, is ultimately without the power to do so. Modern champions of science, however, increasingly emboldened by its discoveries and technological achievements as well as its capacity to institutionalize its triumphs have no such qualms about claiming the ground once occupied by religion.

Though poetry no longer occupies the central place in our cultural life, it did not cede its authority without a fight. From the nineteenth century to the present, there have been recurrent debates with much at stake about what should constitute the core of advanced education. Should it be science or literature (i.e., poetry, the humanities)? What is at stake in answering the question is the formation of the minds and character of the citizenry. The question was the subject of a debate between Matthew Arnold, the chief *apostle* of literary culture (the religious metaphor is not fortuitous), and Thomas Huxley, the champion of Darwinism, the dominant science of the age. On the occasion of the founding of the scientifically oriented Josiah Mason College, Huxley addressed what he believed to be an urgent need for centering education on the sciences rather than literature. The address became the essay "Science and Culture" (1880-81). Huxley begins his essay by noting that the adversaries of a scientific education are not only the literary advocates of a classical education, they are also the practical men of business. He means to disarm "the apostles of culture" (culture here meaning high literary culture) by taking the intellectual high ground. Science, he wants it to be understood, is a disinterested activity not to be reduced to its mechanical applications and utilitarian benefits. And yet it cannot be separated from them. "The distinctive character of our own times lies in the vast and constantly increasing part which is played by natural knowledge. Not only is our daily life shaped by it; not only does the prosperity of millions of men depend upon it, but our theory of life has long been influenced, consciously or unconsciously, by the general conceptions of the universe which have been forced upon us by physical science" (149). (C. P. Snow's argument in "The Two Cultures" [1959] about the necessity of a scientific education and complaint about the scientific ignorance of literary intellectuals is a latter-day version of Huxley's argument.)

In his essay "Literature and Science" (1883), Arnold counters with the claim that, however admirable and important the work of science may be, only literature in its poetic expression can satisfy what he calls "our sense for beauty" and "our sense for conduct." "Following our instinct for intellect and knowledge, we acquire pieces of knowledge; and presently, in the generality of men, there arises the desire for us to relate these pieces of knowledge to our sense for conduct, to our sense for beauty—and there is weariness and dissatisfaction if the desire is balked" (62). This is the heart of the debate. Arnold is the heir of Wordsworth's view of the difference between science and poetry. For all the benefits it brings to mankind, science cannot relate its work to "our sense for conduct" and "our sense for beauty." It should be noted that Arnold, like the scientist, values objectivity in the exercise of poetic imagination and literary criticism. Following Wordsworth, he speaks of "seeing the object as it really is." However, given the differences between the objects of poetic imagination and those of scientific inquiry the ways of objectivity differ. It should also be noted that, as we have already seen, scientists in our own time speak persuasively of their work as having an aesthetic aspect, though it is an aesthetic different from that of poetry, for instance, the symmetry of laws. In the version of poetry that Wordsworth bequeathed to Arnold and the Victorians and the modern world, poetry is man speaking to man. Wordsworth's poetics does not encompass all of poetry. Many poems, particularly modern poems cultivate a sense of difficulty and obscurity that make them as "remote" (Wordsworth's word for science) to readers as any scientific paper.

The debate was conducted with the utmost courtesy; neither Arnold nor Huxley speaks in the accents of triumphalism. Both parties try to find a place for each other's discipline within their schemes of education, though they do not always share a common understanding of each other's discipline. "Literature is a large word," Arnold writes, "it may mean everything written with letters or printed in a book. Euclid's *Elements* and Newton's *Principia* are thus literature" (58). They may be literature, but they are not what Huxley or any practicing scientist means by the modern practice of science. Euclid and Newton belong to the history of science. The practice of science is the exercise of a method or methods, the formulation of hypotheses, the performing of experiments and the making of new discoveries. Scientists don't necessarily have to know the history of their own discipline in order to do their work. Arnold here confuses knowledge of the history of science with the doing of science. Though he dismisses the necessity of a study of the Greek and Roman classics in the original languages, Huxley has a place for modern languages and literatures in his curriculum. The attempt of Arnold and Huxley to conciliate each other should not obscure the fact that what is at stake are different conceptions of what constitutes an educated person in the modern world.

How relevant is the debate to the contemporary scene? Huxley was certainly right to insist on the importance of a scientific education at a time when the dominant cultural elite was hostile to the "increasing part played by natural knowledge" in the world. The contemporary academy has caught up with Huxley, though the increasing specialization and difficulty of scientific study has limited its general appeal to the student body. And Arnold's case for the distinctive aesthetic and moral burden of literary or humanistic study no longer prevails. Arnold's plea for the study of the classics in the original is a lost cause, and Huxley would find that modern literature survives in the academy for the most part in translation. If there is competition between science and the humanities, it has been muted by the contemporary academy, which is ideologically impartial as to the best educational path to take. The modern college curriculum is diverse, and the student not the college makes the choices. It may appear that Huxley has won the debate, for it is the job market that indirectly decides where an institution makes its investments, and the market now favors biology, economics and computer science. The emphases of the curriculum are affected by the biases of the market. Huxley's argument, however, did not rest mainly on the practicality of science, but rather on its intrinsic value as an intellectual discipline. Its practical value was an additional benefit. Whether science as an intellectual discipline has taken hold in the generality of students is another matter. Indeed, there is evidence that it has not taken hold. The physicist and historian of science Gerald Holton, one of the few voices in the wilderness, not only laments the current state of science education in America, but has also proposed practical measures for educating non-scientists in the sciences and thereby overcoming what he calls the silence between science and the humanities.

The association between romanticism and humanism was severed in the early twenties of the previous century, when Irving Babbitt and Paul Elmer More declare a new humanism. Its bias is classical rationalism, its main target Romanticism and its protagonist Rousseau. The new humanism of Babbitt and More became the target of Babbitt's former student T. S. Eliot, who from a religious point of view found humanism guilty of the sin of pride. For Eliot, the universe is God-centered, not human-centered. The American pragmatists William James and John Dewey spoke of themselves as humanists, by which they meant that truths are not absolutes, but rather man made, the result of the exercise of the mind and imagination on reality. They rejected the correspondence theory of truth for its rendering of mind as a passive mirror. The humanism of James, Dewey, and its almost forgotten exponent F.C.S. Schiller may be thought of paradoxically in both the Enlightenment and the romantic tradition.

Tzvetan Todorov's recent work is the most recent sustained effort I know of to revive the humanist tradition, which in his French Enlightenment version includes Rousseau. Though he values the achievements of science, he rejects scientism, for its freedom-depriving attempts to apply the laws of nature to human affairs (social, political and moral life.) Though he is not hostile to religion, he is philosophically indifferent to it. Humanists adhere "to doctrines in which man [and not God] is the point of departure and the point of reference for human actions" (Todorov, 6). Can one speak of a religious humanism, or does the humanist view that "man is the measure" make the phrase an oxymoron? It may be oxymoronic if religion is understood as dogma. But religious *experience* following the understanding of thinkers such as Niebuhr and Jaspers is the struggle of human beings to enhance their humanity through self-transcendence. There seems to me no reason that humanism, generously conceived, cannot accommodate certain claims of religion or religious experience without obliging secular humanists to embrace religion. I am proposing an elasticity to the conception of humanism that might offend those who prefer a stricter, perhaps more systematic definition. My own view of humanism is that it is not a philosophy or an ideology. It is rather a sensibility, a disposition, an attitude, which is responsive to the richness and complexity of human experience. Humanism, in my view, represents a view of life that resists systematic doctrinal definition.

I do not know of many colleagues who think of themselves as humanists in any of the ways I have described above. A few may come close to the nineteenth-century version, which sacralizes the literary tradition. They are in the main practitioners of a discipline: literary criticism or literary history or literary theory or art criticism, etc. The humanistic element is in their appreciation, indeed a love, of the arts. Secular literature becomes in a sense sacred text, the Bible becomes literature, symbolic, not literal truth. I speak of the relatively few in the academy who have remained faithful humanists, because in recent times postmodern skepticism has taken over literary studies. The legitimate complaint against "traditional" humanism was that the appreciation of literature was too often pious and uncritical. But the skepticism directed toward literary works has too often been as indiscriminate and uncritical in their dogmatic skepticism as the appreciators were in their piety.

In an earlier work, *Does Literary Studies Have a Future?* I refer to an ongoing effort on my part to reenfranchise certain ideas in the humanities, "objectivity, disinterestedness, tradition, aesthetic appreciation" (34). I go on to say that "their loss of credit is a misfortune for the academy," and that they "should be the common possession of scholars of whatever persuasion, left, right or center." In the present context, I would say that these virtues, particularly those of objectivity and disinterestedness, should be the common possession of both scientists and humanists. The differences between the sciences and the humanities are no cause for quarrels between them. My quarrel with the neo-Darwinists is that they have committed an ideology, not a science in their failure to be disinterested and objective in their approach to the disciplines of the humanities. In a time like the present when the interdisciplinary is the rage, one should be aware of how it can become a vehicle of reductionism, the impoverishing translation of one field into another. A discipline has its own imperatives, which may or may not cross or converge with another discipline. Unless the interdisciplinary is based on mutual respect among disciplines and a sense of its own limits, interdisciplinary work becomes vacuous. The prospect of disciplines going off in various directions, unconstrained by the demand for consilience may bring greater rewards than the opposite and illusory prospect of the unity of all knowledge, so tantalizing to neo-Darwinists.

Works Cited

American Scholar, Summer, 2005

Anderson, Benedict. *Imagined Communities: Reflections on the Origin and Spread of Nationalism*. London: Verso, 1983.

Appleby, Joyce, Hunt, Lynn; Jacob, Margaret. *Telling the Truth about History*. New York: W. W. Norton, 1994.

Arnold, Matthew, "The Study of Poetry", *Complete Prose Works* vol. ix, ed. R.H. Super. Ann Arbor: University of Michigan, 1960.

—— "Literature and Science", *Complete Prose Works* vol. x, ed. R. H. Super. Ann Arbor: University of Michigan Press, 1960.

Barash, David P. and Barash, Nanelle. *Madame Bovary's Ovaries: A Darwinian Look at Literature*. New York: Delacorte Press (Random House), 2005.

Barzun, Jacques. *Darwin, Marx, Wagner: Critique of a Heritage* (1941) New York: Doubleday Anchor Books, 1958.

A Stroll with Willam James, New York: Harper & Row, 1953.

Beer, Gillian. *Darwin's Plots: Evolutionary Narrative in Darwin, George Eliot and Nineteenth Century Fiction*. London: Routledge and Kegan, 1983.

Berlin, Isaiah. *The Sense of Reality: Studies in Ideas and their History*. New York: Random House (Pimlico), 1996.

Bloch, Marc. *The Historian's Craft*. Tran. Peter Putnam. New York: Alfred A. Knopf, 1953.

Carlyle, Thomas. "Signs of the Times" *The Works of Thomas Carlyle in 30 vols.* Ed. H. D. Traill, vol. xxvii. New York: AMS Press, 1974.

Carr, E.H. *What Is History?*. New York.: Random House (Vintage). 1961.

Casti, John L. *Paradigms Lost*. New York: William Morrow, 1989.

Carroll, Joseph, *Evolution and Literary Theory* Columbia: University of Missouri Press, 1995.

——*Literary Darwinism: Evolution, Human Nature and Literature*. New York and London: Routledge, 2004.

Coetzee, J. M. *Elizabeth Costello*. New York: Viking Press, 2003.

Culler, Dwight. "The Darwinian Revolution and Literary Form" in Appelman, Phillip ed. *Darwin*. Norton Critical Edition, 1961.

Dyson, Freeman. New York Review of Books, May 13, 2004.

Diamond, Jared. *Guns, Steel, and Germs: The Fates of Human Societies*. New York: W.W. Norton & Company, 1999.

Dawkins, Richard. *The Selfish Gene*. New York: Oxford University Press, 1976.

——*Unweaving The Rainbow: Science, Delusion and the*

——*Appetite for Wonder*. Boston and New York: Houghton Mifflin, 1998.

——"The Atheist" Interview with Gordy Slack, Salon.com April 25, 2005.

——*The God Delusion*. Boston: Houghton-Mifflin, 2006.

Dennett, Daniel C. *Consciousness Explained*. Boston: Little Brown, 1991.

―――*Darwin's Dangerous Idea: Evolution and the Meanings of Life*. New York: Simon and Shuster, 1995.
―――*Breaking the Spell: Religion as a Natural Phenomenon*. New York: Viking Press, 2006.
Feynman, Richard. *The Meaning of it All*. Reading, MA: Perseus Books, 1998.
―――*The Pleasure of Finding Things Out*. Reading, MA: Perseus Books, 1999.
Fleming, Donald. "Charles Darwin, The Anaesthetic Man in Appleman, Phillip, *Darwin*, ed. New York: The Norton Critical Edition, 1961.
Foucault, Michel. "What is an Author" in Rabinow, Paul. Ed. The Foucault Reader. New York: Pantheon Books, 1994
Galison, Peter and Stump, David J., Eds. *The Disunity of Science: Boundaries, Contexts and Power*. Stanford: Stanford University Press, 1996.
Gleick, James. *Chaos: Making a New Science*. New York: Viking Penguin, 1987.
Goodheart, Eugene. *The Reign of Ideology*. New York: Columbia University Press, 1996.
―――*Does Literary Studies Have a Future?* Madison: University of Wisconsin, 1999.
Gould, Stephen Jay. *The Hedgehog, the Fox and the Magister's Pox: Mending the Gap between Science and the Humanities*. New York: Harmony Books, 2003.
Greene, Brian. "The Time We Thought We Knew," *New York Times*, January 1, 2004.
Gross Paul and Levitt, Norman. *The Higher Superstition: The Academic Left and its Quarrel with Science*. Baltimore and London: The Johns Hopkins University Press, 1984.
Haack, Susan. *Defending Science – Within Reason: Between Scientism and Cynicism*. Amherst, N.Y.: Prometheus Books, 2003.
Hacking, Ian. *Representing and Intervening*. Cambridge, England: Cambridge University Press, 1983.
―――Ed. *Scientific Revolutions*. Oxford: Oxford University Press, 1981
―――Review of Steven Rose's *The 21st Century Brain Explaining, Mending and Manipulating the Mind*. *London Review of Books*, 18/8/05
Helmholtz, Herman von. *Science and Culture*. Chicago: The University of Chicago Press, 1995.
Holton, Gerald. *Thematic Origins of Scientific Thought: Kepler to Einstein*. Cambridge, MA: Harvard University Press, 1988.
―――*Victory and Vexation: in Science: Einstein, Bohr, Heisenberg and Others*. Cambridge, MA: Harvard University Press, 2005.
Himmelfarb, Gertrude. *Darwin and the Darwinian Revolution*. New York: Doubleday Day Anchor Books, 1959.
Huxley, Julian S. *Man Stands Alone*. New York and London: Harper & Brothers, 1941.
Huxley, Thomas. "Science and Culture". *Collected Essays* vol. iii. New York: Greenwood Press, 1968.
―――*Evolution and Ethics and other essays*. New York: D. Appleton and Company, 1899.
James, William, *Pragmatism and Four Essay from the Meaning of Truth*. New York: New American Library, 1974.
Johnson, George. "A Free-for-All on Science and Religion". New York Times, November 21, 2006 D6.
Kimball, Roger. "Fallible but not Futile" Times Literary Supplement, January 23, 2004.
Kelley, Aileen, *View from the Other Shore: Essays on Herzen, Chekhov and Bakhtin*. New Haven, CT: Yale University Press, 1999.
Koestler, Arthur. *The Ghost in the Machine*. London: Hutchinson, 1967, '76.

Kuhn, Thomas. *The Structure of Scientific Revolutions*. Chicago: The University of Chicago Press, 1962.
——*The Road Since Structure*. Eds. James Conant James Haugeland. Chicago: University of Chicago Press, 2000.
Leavis, F.R. "Two Cultures? The Significance of C.P. Snow" New York: Pantheon Books, 1962.
Lewontin, Richard; Rose Steven; Kamin Leon J. *Not in Our Genes: Biology, Ideology and Human Nature*. New York: Pantheon, 1984.
Max, D. T, "Literary Darwinism". Sunday Magazine Section, New York Times. November, 8, 2005.
Midgley, Mary. *Science and Poetry*. London: Routledge, 2001.
Niebuhr, Reinhold. *Nature and Destiny of Man: A Christian Interpretation*. New York: Charles Scribner's Sons, 1949.
Patai, Daphne and Corral, Will. *Theory's Empire*: *A Dissenting Anthology*. New York: Columbia University Press, 2005.
Pater, Walter. *Appreciations*. London: Macmillan, 1924.
Pinker, Steven. *The Blank Slate: The Modern Denial of Human Nature*. New York: Viking, 2002.
Plumb, J. H. *In the Light of History*. New York: Dell (Delta Books). 1972.
Popper, Sir Karl Raimund, *The Poverty of Historicism (1957)*. New York: Routledge, 1988.
Ringer, Fritz. *Max Weber: An Intellectual Biography*. Chicago: The University of Chicago Press, 2004.
Ruse, Michael. *The Evolutionist-Creationist Struggle*. Cambridge. MA: Harvard University Press, 2005.
Singer, Peter. *Rethinking Life and Death: The Collapse of Our Traditional Ethics*. New York: St. Martin's Griffin, 1994, '96.
Snow, C.P. *The Two Cultures* (1959, '63) Cambridge: Cambridge University Press, 1998.
Storey, Robert. *Mimesis and the Human Animal: on the Biogenetic Foundations of Literary Representation*. Evanston, IL. Northwestern University Press, 1996.
Todorov, Tzvetan. *Imperfect Garden: The Legacy of Humanism*. Trans. Carol Cosman. Princeton, NJ: Princeton University Press, 2002.
Weinberg, Steven. *Dreams of a Final Theory: The Scientist's Search for the Ultimate Laws of Nature*. New York: Vintage Books, 1994.
——*Facing Up: Science and Its Cultural Adversaries*. Cambridge, MA. 2003.
Whitehead, Alfred North. *Science and the Modern World* (1925) (New York: The Macmillan Company, 1957).
Wilson, E.O. *Consilience: The Unity of Knowledge* (New York: Vintage Books, 1998).
Wordsworth, William. *Preface* and *Appendix* to *Lyrical Ballads, Wordworth's Literary Criticicsm* Ed. W. J. B. London and Boston: Routledge and Kegan Paul, 1974.

Index

Aeschylus, 36
Anderson, Benedict, 27-28, 29
Appleby, Joyce, 74, 75
Aristotle, 34, 90, 106, 111
Arnold, Matthew, 20, 111-112, 115, 116-117
Augustine, St., 34
Austen, Jane, 21-22
Ayer, A.J., 94

Babbitt, Irving, 118
Bach, Johann Sebastian, 38
Balzac, Honoré de, 13
Barash, David P., 11-14, 19, 40
Barash, Nanelle, 11-14, 19, 40
Barzun, Jacques, 80-82
Baudelaire, Charles, 18
Beer, Gillian, 13
Bergson, Henri, 80, 81
Berlin, Isaiah, 71, 72, 90-91, 92, 96
Bierce, Ambrose, 19
Blake, William, 17, 112, 114
Bloch, Marc, 67, 71-72
Bloom, Haqrold, 105
Bohr, Niels, 100, 101, 102
Brown, Joe E., 92
Burke, Edmund, 73
Bush, George W., 51
Butler, Samuel, 80-81

Camus, Albert, 42-44
Carlyle, Thomas, 114, 115
Carnap, Rudolph, 44
Carr, E.H., 66-73, 75
Carroll, Joseph, 19-21
Casti, John, 3, 106
Charlemagne, 70
Chomsky, Noam, 7
Coetzee, J.M., 59-60
Coleridge, Samuel Taylor, 20, 106
Conrad, Joseph, 86
Copernicus, 51, 57

Culler, Dwight, 22

Dante, 38, 95
Darwin, Charles/Darwinism, 13, 19, 21, 24, 29, 31, 38, 39, 51, 54, 55-56, 57, 58, 59, 60, 61, 62, 63, 66, 77, 80, 81-82, 109, 111
Darwin, Erasmus, 81, 82
Dawkins, Richard, 2-3, 28-36, 38, 39, 46, 54, 61, 92, 95, 102, 106
De Kooning, Willem, 15
Dennett, Daniel, 2, 3, 24, 39-42, 61-63, 92
Descartes, René, 5, 111
Dewey, John, 118
Diamond, Jared, 65-66
Dickens, Charles, 83, 114
Donne, John, 36
Donoghue, Denis, 23
Dostoevsky, Fyodor, 13, 38, 87, 95
Douglas, Mary, 32, 39
Dryden, John, 20
Durkheim, Emile, 37
Dutton, Dennis, 14, 17
Dyson, Freeman, 102-103

Easterlin, Nancy, 19
Einstein, Albert, 16, 29, 44, 100, 103
Eliot, George, 107
Eliot, T.S., 16, 23, 24-25, 106, 115
Empson, William, 20
Erasmus, 110
Euclid, 117
Evans-Pritchard, Edward, 39-40

Feyerabend, Paul, 109-110
Feynman, Richard, 104-105
Flaubert, Gustave, 12, 18
Ford, Henry, 33, 65
Foucault, Michel, 62
Frankenthaler, Helen, 15
Freud, Sigmund, 12, 13, 16, 31, 62

Frye, Northrup, 23

Galileo, 62, 111
Gibbon, Edward, 73, 76
Gish, Duane T., 79, 80
Goethe, Johann Wofgang von, 113-114
Goodwin, Brian, 84
Gottschalk, Jonathan, 22
Gould, Stephen J., 4, 6, 62, 79
Graubard, Stephen, 46
Greene, Brian, 42-46, 100, 102, 103
Gross, Paul, 93-95

Haack, Susan, 35-36, 97
Hacking, Ian, 9-10, 37
Handel, George Frederick, 38
Hartman, Geoffrey, 23
Hawking, Stephen, 58
Hayek, F.R., 10
Hegel, G.F.W., 69, 87
Heisenberg, Werner, 16, 93, 103
Helmholtz, Hermann von, 113
Herbert, George, 36
Himmelfarb, Gertrude, 79, 80, 82-83
Hitler, Adolf, 70
Hoffding, Harald, 101
Holton, Gerald, 100, 101, 118
Hume, David, 2, 73
Hunt, Lynn, 74, 75
Huxley, Julian, 28
Huxley, Thomas, 55, 115-116, 117
Jacob, Margaret, 74, 75
James, Henry, 17
James, William, 38, 57, 101, 118
Jaspers, Karl, 118
Johnson, Samuel, 20
Joyce, James, 16, 115
Judd, Donald, 15

Kahler, Erich, 111
Kamin, Leon J., 6
Kant, Immanuel, 5, 60, 61
Keats, John, 24, 83
Kepler, Johannes, 111
Kierkegaard, Søren, 101-102
Kilmer, Joyce, 24
Kimball, Roger, 97
Kline, Franz, 15
Koestler, Arthur, 92
Kolakowski, Leszek, 30
Kuhn, Thomas, 98, 99-100, 106-107

Lakatos, Imre, 98
Lawrence, D.H., 32
Leach, Edmund, 32
Leach, Edmund, 39
Leavis, F.R., 20, 23
Leibnitz, 47, 67
Lemmon, Jack, 92
Lenin, Vladimir, 70, 71
Levitt, Norman, 93-95
Lewontin, Richard, 6-9
Lindbergh, Charles, 70
Locke, John, 2, 112
Louis, Morris, 15

Mann, Thomas, 107
Marx, Karl/Marxism, 16, 31, 62, 69, 90
Max, D.T., 21-22
McEwen, Ian, 107
Medawar, Peter, 46, 114
Michelangelo, 38
Midgely, Mary, 83-86
Mill, James, 83
Mill, John Stuart, 61, 83, 98
Milosz, Czeslaw, 30
Milton, John, 13, 29, 38, 48
Montaigne, Michel de, 34
More, Paul Elmer, 118
Morris, William 114

Nabokov, Vladimir, 17
Nagel, Thomas, 1
Napoleon, 70
Newman, Barnett, 15
Newton, Isaac, 44, 62, 111, 112, 113, 114, 117
Niebuhr, Reinhold, 35, 54-55, 118
Nietzsche, Friedrich, 8, 58, 81
Noland, Kenneth, 15

O'Casey, Sean, 30
Olitski, Jules, 15
Orwell, George, 110

Paine, Thomas, 111
Parkman, Francis, 73
Pater, Walter, 115
Penrose, Roger, 25
Pinker, Steven, 14-19, 20, 95
Plantinga, Alvin, 79, 80
Plato, 34, 45
Plumb, J.H., 73
Polanyi, Michael, 24

Pollock, Jackson, 15
Popper, Karl, 10, 46, 67, 69, 71, 72, 76-77, 85, 97-99, 102
Pound, Ezra, 16
Putnam, Hilary, 98-99

Ransom, John Crowe, 20
Raphael, 38
Richards, I.A., 20
Ricks, Christopher, 23
Ringer, Fritz, 47, 72
Roosevelt, Franklin, D., 70
Rose, Stephen, 37
Rose, Steven, 6
Roth, Philip, 70
Rothko, Mark, 15
Rousseau, Jean Jacques, 34, 118
Ruse, Michael, 36, 38, 79-80
Ruskin, John, 114
Rutherford, Ernest, 103

Schiller, F.C.S., 118
Schrodinger, Erwin, 103
Shakespeare, William, 13, 38, 48, 95
Shattuck, Roger, 23
Shaw, George Bernard, 80, 81
Shelley, Percy Bysshe, 113
Simpson, G.G., 33
Singer, Peter, 51-61,
Skinner, B.F., 7, 12, 61
Snow, C.P., 74
Snow, C.P., 74, 105, 116
Sokal, Alan, 93
Spencer, Herbert, 58, 90

Spinoza, Benedict, 111
Stella, Frank, 15
Stendhal, 18
Stevens, Wallace, 20
Still, Clyfford, 15
Storey, Robert, 19
Swinburne, Richard, 36
Szymborska, Wislawa, 60

Thucydides, 73, 76
Tillich, Paul, 36
Todorov, Tzvetan, 10, 118
Tolstoy, Leo, 38, 45-46, 90, 95
Trilling, Lionel, 20

Voltaire, 47

Watson, John, 12
Weber, Max, 45, 47, 72
Weinberg, Steven, 5-6, 85, 89-92, 93, 97, 99-100, 102, 103
Wells, H.G., 86
Whewell, William, 4
White, Hayden, 75
Whitehead, Alfred North, 114
Williams, George, 31
Williams, Raymond, 20
Wilson, Edward O., 1-2, 3, 4-5, 6, 8, 9, 11, 13, 22, 38, 55, 61, 66, 83, 84, 95, 102
Wittgenstein, Ludwig, 79
Wolfe, Tom, 15, 17
Woolf, Virginia, 14
Wordsworth, William, 83, 112-113, 116-117